WHAT MAKES YOU WHAT YOU ARE

A FIRST LOOK AT GENETICS

WHAT MAKES YOU WHAT YOU WHAT YOU ARE

A FIRST LOOK AT GENETICS

SANDY BORNSTEIN

JULIAN MESSNER

Text copyright © 1989 by Sandy Bornstein
Illustrations copyright © 1989 by Frank Cecala

Cover photos: left copyright © Adam Smith/Westlight;
right, top and bottom copyright © 1989 Agatha Lorenzo

JULIAN MESSNER and colophon are trademarks of Simon & Schuster, Inc.
Design by R STUDIO T • Raúl Rodríguez, Rebecca Tachna.
Manufactured in the United States of America.

The material in Activity Box Two, p. 33, is adapted from and reproduced with
permission from *Science and Children,* Sept. 1987. Copyright 1987 by the
National Science Teachers Association, 1742 Connecticut Avenue, NW,
Washington, DC 20009.

Library ed. 10 9 8 7 6 5 4 3 2 1
Paper ed. 10 9 8 7 6 5 4 3 2 1

Library of Congress Cataloging-in-Publication Data

Bornstein, Sandy.
What makes you what you are : a first look at genetics / Sandy Bornstein.
p. cm.
Bibliography: p.106
Includes index.
Summary: Discusses the science of genetics and what it has revealed about
how genes determine the blueprint for life.
1. Genetics—Juvenile literature. [1. Genetics.] I. Title.
QH437.B68 1989
575.1—dc20   89-9440
CIP
AC

ISBN 0-671-63711-8   ISBN 0-671-68650-X (pbk.)

FOR MY OWN FAMILY TREE IN
JOYFUL AND LOVING
CELEBRATION

CONTENTS

INTRODUCTION
"WHY, YOU LOOK JUST LIKE…"
1
CHAPTER 1
FAMILY TIES
5
CHAPTER 2
CELLS, CELLS, EVERYWHERE
13
CHAPTER 3
CELL DIVISION — MANY FROM ONE
23
CHAPTER 4
THE NEXT GENERATION
35
CHAPTER 5
IN THE SHADE OF YOUR FAMILY TREE
51
CHAPTER 6
HUNTING FOR THE BLUEPRINT
70
CHAPTER 7
THE BUSY RESTING CELL
80
CHAPTER 8
FINDING THE CONNECTIONS
92
BIBLIOGRAPHY
106
INDEX
109

INTRODUCTION

"WHY, YOU
LOOK JUST
LIKE...

"My goodness! You look just like . . ." You must have heard this dozens of times. The very first time it was said about you, you were probably too young to know what was happening. Within the first few days of your life, perhaps even in the first few moments, people wanted to know whom you looked like in your family.

It always seems to happen at those large family reunions. The first thing your relatives usually notice is how big you have grown since the last time they saw you, but it doesn't stop at that. Sooner or later you will hear someone say, "You look more and more like your [mom, dad, sister, brother, Aunt Lil] every year!" Sometimes you "look just like" two or three *different* people depending on which relative is talking to you.

It's quite natural for people to make comparisons like this. It seems every time we look around us, especially when we are examining something closely, we are really looking for similarities and differences. We're always looking for patterns—things that look alike, things that behave alike. Of course, we do that with people too.

When your relatives look at you and try to figure out whom you look like, it probably seems to them that you're not *entirely* like just one other person. Sometimes it seems that you have bits and pieces of many of your relatives. Perhaps you have your uncle Charlie's ears, Grandma Anna's nose, and your cousin Arthur's tiny feet.

All of these features are called characteristics, or *traits*. Traits include all sorts of things—hair color, skin tone, the shape of your ears, the size of your hands and feet, your height, and your weight. You might also include more subtle features like mannerisms, talents, and interests. You may have been told that you walk, talk, or laugh like someone in your family. You may even have been told that you've got someone's interest in music, talent for carpentry, or passion for stargazing.

Other people aren't the only ones who notice similarities. You may have noticed many family resemblances yourself. As you look through old family photos, you may see the shape of your face, the cleft in your chin, or the dimple when you smile right there in the features of a distant relative. When you see resemblances that are so obvious in a family, it may seem that these traits are passed along from one relative to another, that they are *inherited*.

How much of what you are is *really* inherited from your family? Can height, eye color, ear shape, or artistic talent be passed from generation to generation like a precious family heirloom? Can people ever choose what they give or what they get?

FIGURE 1 *Where do you come from?* Your traits may be traced to many members of your family.

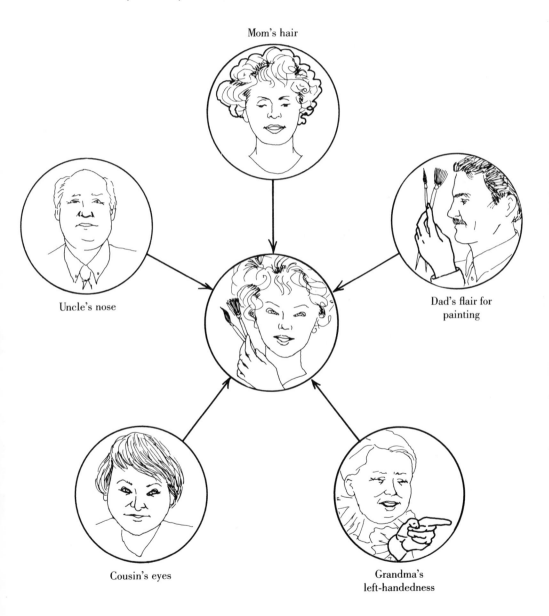

Mom's hair

Uncle's nose

Dad's flair for painting

Cousin's eyes

Grandma's left-handedness

Some people are trying to find answers to these questions. We know that what is inherited is not the trait itself. (Cousin Arthur doesn't actually *give* you his tiny feet, or what would he have to stand on?) Instead, we get information—a set of instructions—from our relatives that directs how we grow. The science that studies how this inheritance guides our growth and development is called *genetics*. It is also the study of how that information came to us from our parents. Still, we are all more than just what we inherit. Genetics also studies how the environment that we live in, even the one inside our mothers before we are born, can influence and change how we turn out.

As a science, genetics is very new, but it is based on a human interest that is very old. People have been gathering information about inheritance for many thousands of years. It seems we have always been curious about why people turn out the way they do.

Yet, despite all this interest, a regular, careful, "scientific" study of the way traits are passed from parents to children was not begun until a little over a hundred years ago. Our understanding of some of the actual mechanics of how inheritance works is less than forty years old. The ancient puzzle of why we are what we are is still unsolved, but new pieces to this puzzle are being found all the time.

In this book we will look at some of the things that the science of genetics has discovered and some of the questions that are still unanswered. We will look at what we know about family resemblances. We will look at what we're made of and how we grow. We will look at the powerful chemicals that help control the way we develop. We will look at the many similarities between living things, and at the forces that let them be so very different.

CHAPTER 1

FAMILY TIES

Looking back at earth from out in space, astronauts say it looks like a big blue marble. From way up there you can see oceans and continents, hurricanes and, at night, the lights of big cities. You can't see all the living creatures that inhabit this planet, but there is an incredible number of them—far too many to count in your lifetime. In fact, if you could arrange to count *just* the human beings on the planet, and if you counted at the rate of one person per second it would take you over 126 years to count the ones that are here right this moment! That doesn't include all the babies that would be born while you were counting.

Over four billion people and all of them different—yet are they really *that* different? We share many features just because we are human beings. You know what makes you human, but if someone asked you for a description of a human being, what traits would you include? You probably wouldn't have much trouble coming up with a long list of characteristics.

You might start your list with common body parts that most of us have—a head with two eyes, two ears, a nose, and a mouth; arms with hands and fingers that can grab; legs with feet and toes. Even our insides look an awful lot alike, with a hard inside skeleton, four-chambered heart, large brain, paired lungs and kidneys, stomach, intestine, liver, and so forth. You might include our five senses of touch, taste, smell, sight, and hearing, which tell us what is going on in the world around us. You could add that our babies are born alive instead of hatching from eggs. We are social animals that live in groups. We make and use tools. We care for our children for a long time after they are born, teaching them many things so they can care for themselves. The list of common traits could really go on and on. Not every person has to have all of these features, and the traits don't all have to work equally well or look the same in every individual. The fact that humans have so many features in common, and the fact that humans can produce children only with other humans, makes us members of the same *species*.

Even with all these similarities, we know that no two humans are exactly alike. Even identical twins have different traits. We share many human features with all other people, yet there is still plenty of room for variation. For example, there is an average height for an adult person, yet the human species includes people who are under three feet tall and people who are almost eight feet tall.

People are communicating animals, yet their voices are of many tones and they make many different sounds, signs, gestures, body movements, and facial expressions to get their messages across. Most humans have hair, yet think of the variety. There are several colors, dozens of shades, and texture that may be anywhere from coarse to fine. Hair may be straight or wavy or curly, and a person may have any amount—from a thick head of hair to a sparse covering. Hundreds of combinations are possible just for hair! No wonder we can say no two people are exactly alike.

Still, we said people's looks are not accidental, and the specific features *you* have are not accidental, either. You *do* look more like the members of your *own* family than like the family of your best friend. You probably don't look like just *one* member of your family, but like some combination—a mixture of features from your mother's side and your father's side of the family.

How you look is controlled by rules that might seem to contradict each other. First, because you are a member of the human species, you look like other humans. We wouldn't confuse a member of our own species with a chipmunk or a swan or a pine tree. Second, because so much variety is possible in those features that make you a human being, you are unique, different from all other human beings. Third, although you are different from other people, some people are more like you than others, and these are members of your family. Your relatives have *specific* features in common with one another. Instead of hundreds of possibilities for each feature, a family may have only a few. There are still many ways these limited possibilities can be combined and recombined. That is why family members look similar but not identical.

These principles don't just apply to people. They also hold true for the entire community of living things— the millions of other plant and animal species with whom we share the planet. If you have ever been lucky enough to see a complete litter of animals—like hamsters, kittens, or puppies—then you know how strong family resemblances can be.

You also know how the "mixing up" of family features can make for a lot of variety. Each member of the litter is unique. There are always differences in size, coloring, shape, and combination of features. Sometimes there is a giant, and sometimes there is a runt in the litter. Even with their differences, all of the brothers and sisters look a lot alike. They resemble one another because they share the features of their mother and their father.

Although we don't choose our own traits or the traits for our babies, we do sometimes choose traits for the animals and plants we depend upon for food, work, and pleasure. We are able to do that by carefully choosing who will be the mothers and fathers of the animals and plants we want to breed or grow.

When two mixed-breed dogs produce a litter of puppies, their babies, or *offspring*, may look a little like each parent. If the parents don't look like each other, there may be a lot of different-looking features for the puppies to share and to mix up in different combinations. One parent might have long ears and the other short ones. The puppies might have pointy ears or floppy, long hair or short, dark fur or spots, curled tails or straight; the possibilities are almost endless. The more different the parents are from each other, the more different features there can be in the babies (see figure 2).

If breeders want to produce dogs with specific features, they have to pick the parents carefully. By taking two dogs who have the features that they want, and by using them to produce their next litter of puppies, breeders have developed over many generations more than 125 pure breeds of dogs. They are all members of the dog species, yet range in size and appearance from the tiny Chihuahuas and toy Manchester terriers to the huge Great Danes and Saint Bernards. Puppies

FIGURE 2   *Sharing family features.*

a. If the parents are very different, the offspring mix and combine many traits to create variety.

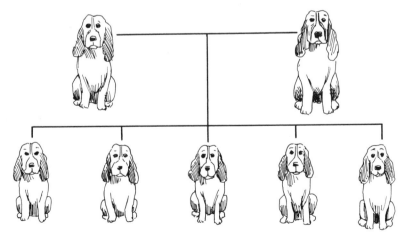

b. If the parents are very similar, the offspring have fewer different combinations. Everyone looks alike.

from purebred dogs look very much like their parents, who look like each other. There aren't so many different possible combinations of features, so all members of a breed tend to look a great deal like one another.

We say that parents pass their traits on to their offspring, but how can this be done? In ancient Greece, people believed that there was a fluid that moved through the parent's body, that "remembered" in some way the actual shape of the parent's features. When the fluids from two parents were combined, their features were blended in their offspring.

The Greeks were right that *something* was carrying information from each parent to the child. The features themselves are not passed on from parent to child, however. Instead, there are instructions on how to recreate these features in the offspring. The instructions are carried in a special cell from each parent. These special cells combine to produce the new organism. We now call these instructions *genes,* and this is where the term "genetics" comes from.

For a while science didn't really move so far away from the Greek idea of inheritance. Even in the early years of modern genetics, it was argued that every feature, or trait, that an organism possessed was controlled by a single pair of genes, one gene of the pair coming from each parent. All of the gene pairs added together could create the new organism. In people, there would be one gene pair for eye color, one for nose shape, another for height, still another for foot size, and so forth. The genes could be read off feature by feature to put together a new individual.

Now we know that the instructions are more complicated than that. Most traits are influenced by the action of more than one gene. There are genes that may have impact on more than one trait. All of your genes work together to control your development. You get your genes from both of your parents. It is the genes that help determine what you will look like and how you will develop. Development is a process. As the new organism grows, all of the developing parts have an impact on one another. It is not just a matter of snapping together the parts.

Even without knowing about genes, farmers have been using the inheritance of features for thousands of years to improve their stock. We all know that cows always produce calves, horses always produces foals, and chickens always produce chicks. But cows are not all identical to one another, nor are horses, nor are chickens. Just as people are different from one another, so each cow, each horse, each chicken is a unique individual. Some cows may be bigger, meatier, or better milk producers than other members of their herd. By breeding the cows with the features they want most, farmers can improve the next generation. With careful selection, the milk production or meat production of the entire herd can be increased.

Breeding programs aren't limited to food-producing animals. They are also used with plants. The seeds of a pea plant will always produce more pea plants, but the new plants may not all be identical. Some may produce sweeter peas or more peas, or may grow faster than other plants. Breeders pick the seeds for next year's crop from those plants that produce the *best* fruits and vegetables, from the plants that grow fastest, or from the plants that best resist disease, hoping to keep these traits and develop stronger, healthier plants.

Carefully picking the parents of the next generation to "improve" the characteristics of the animals and plants we raise is called *artificial selection*. It does not occur naturally; it is a human invention. It started with very detailed observations of how living creatures are alike and how they are different, with the observers always looking for patterns.

The living world is filled with these similarities and differences. They have fascinated people for ages. We will explore some of the most basic similarities in the next chapter.

CHAPTER 2

CELLS,

CELLS,

EVERYWHERE

Take a walk through the park in late spring. Everywhere you look there are signs of life. The trees are covered with leaves that have just recently appeared. They are changing from the light yellow-green of their first days to the deeper shades we will see all summer. Squirrels run across the lawns looking for food. Early-blooming dandelions have changed their bright yellow flowers for the delicate puff-balls of "wish" seeds that scatter in the breeze. If you look and listen carefully, you might find a nest of baby birds fresh out of their eggs. Tadpoles swim below the surface of the pond, while newly emerged dragonflies buzz above the water.

All of these creatures are busy with the struggle to stay alive. They all need energy to live. They seek food from their surroundings. They use this nourishment to grow and change; they discard what they cannot use. If they are injured, they use energy to repair themselves. They must use a part of their energy searching for more food and more energy. At some time in their lives, they produce young very much like themselves, who follow the same cycle of life as they did, and as their parents did before them.

There is so much variety among these living things! Some are plants and some are animals. Some are huge and some are barely visible. Some are rooted to the ground while others can run, fly, or swim. Some are covered with rough bark while others are covered with fur or feathers. They come in many different shapes and colors. Each of these individual living beings is called an *organism*.

You wouldn't have trouble distinguishing between the squirrel hunting for acorns on the lawn and the robin hunting for worms. You wouldn't confuse the dandelion with the giant evergreen. After all, they are so different! Yet all of these—in fact, all living things—are more alike than you might suspect. If you could look at them closely, if you could magnify them many times their normal size, you would find they are all made of the same tiny building blocks called *cells*.

There are, of course, some differences in the size and shape of these building blocks. The *most important* differences between living organisms are in the number and arrangement of the blocks, rather than in the basic materials they're made from. When people construct a building, they can use the same materials, bricks or cinder blocks, to make totally different buildings. They can build a small house or a large hospital just by using a different blueprint or set of architect's plans and by arranging the bricks in different ways. Similarly, living organisms follow different "blueprints" as they grow and develop, but the building materials they work with are startlingly alike.

Even between plants and animals, there are

more similarities than differences between the building blocks, or cells.

How do we know about these tiny cells? What do these "building blocks" look like? Can you see cells? Not usually. They are generally so small that you would have to enlarge them hundreds of times to see them. There is, however, a large cell that you have probably seen many times and not realized that it was a cell. The yolk of a hen's egg is a single cell from which (under the right conditions) an entire new chicken can grow. It is large and swollen because it has lots of food stored inside it for the developing chick. It is swollen so big that it is easy for us to see.

But how do we know about other cells, the ones that are too small for the human eye to see? Until a few hundred years ago science knew nothing at all about cells. Their discovery took place after the invention of a device that magnifies objects to many times their normal size. The result is that much more detail can be seen. This invention was called the *microscope*.

Before the invention of the microscope about three hundred fifty years ago, the study of living things was limited to what could be seen with the human eye. Sharp, careful observations had revealed a lot about animals and plants. Scientists had looked at living creatures on the outside and inside. But how could they know anything about cells, which are too small to feel and too small to see? The fact is they couldn't.

When the microscope was invented, it created a great stir among scientists. The machine was an arrangement of carefully ground pieces of glass, called *lenses*, and a tube to hold them in place. It sounds so simple, yet this device opened

LOOKING AT A VERY LARGE CELL

YOU WILL NEED   a raw egg
a small bowl
a spoon
a magnifying lens, if you have one

WHAT TO DO   Hold the egg over the bowl. Very gently crack the egg by tapping it with the edge of the spoon. Make the crack in a ring around the middle of the shell. Split the eggshell with your thumbs and carefully pull the halves of the shell apart. Spill the unbroken egg into the bowl and examine it carefully.

This is what the egg looks like *in the shell.* See if you can find these parts in the opened egg.

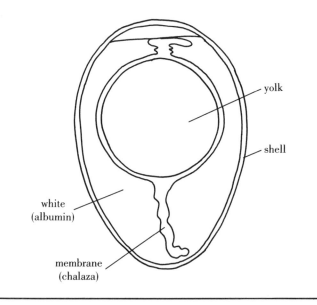

yolk

shell

white
(albumin)

membrane
(chalaza)

up a world of study that had never before been imagined.

People began to look at everything they could think of under the microscope. They looked at living things and nonliving objects. Everything looked so different, so big, so clear, so interesting. All of the curiosity that had driven people to study the visible world with such care was now turned to the new world revealed by the microscope.

One of the scientists who was fascinated by the microscope was named Robert Hooke. He examined many things under his microscope, among them, a very thin slice of a cork tree. He saw tiny compartments. To Hooke they looked like tiny, empty rooms all pressed together, like the ones in which monks lived in monasteries, so he named them *cells*. Actually, he was looking at the walls left behind after the building blocks of the tree died. The name "cell" stuck, and it is the word we use today.

People found cells in all of the living materials they examined. Pieces of plants, the tissues of animals, blood samples—they were all made of cells. They even looked at samples of pond water. In them, the microscope revealed a tiny world filled with many kinds of little beasts. Lots of them were made of only one cell. Others were made of many cells arranged together. No one had ever suspected that in a single drop of pond water there was so much life, so much variety! Hundreds of creatures swam about hunting for food, sometimes being hunted themselves by larger but still microscopic organisms.

Looking at these tiny worlds became such a popular pastime that some people bought microscopes to keep in their homes to entertain their guests at dinner parties. And everywhere people looked in the living world there were cells. Here was another pattern to be explained! Scientists took all of

17

these widespread observations and developed something called "the cell theory"—the idea that *all* living things are composed of cells. They might be different sizes, have different shapes, or do different jobs, but all living things are made of cells.

So, what is a cell like?

Every cell has an outer "skin" called a *membrane*. This serves as a container for the oozy liquid inside the cell. Although the membrane isn't hard, it gives some shape to the cell, limits its size, and keeps stuff from dripping out.

Since it is the outer edge of the cell, the membrane is what comes in contact with the "world outside." Sometimes its job is to serve as a barrier, keeping things inside the cell from leaking out all over the place, preventing things on the outside of the cell from coming in. But don't get the idea that nothing at all can pass through the membrane; if that were so, the cell would be isolated from the world around it. Where would it get food? How would it get rid of its wastes? So, sometimes the membrane works like a delivery system, bringing in small particles the cell needs, passing to the outside things the cell needs to dump (see figure 3).

The membrane is also a source of information to the cell. Just as your skin tells you about the world around you (if it is hot or cold, if things you touch are sharp or smooth), the cell membrane must alert the cell to changes in its surroundings. In tiny organisms, the cell membrane may even locate food or help avoid dangers.

Inside the cell membrane is a jellylike substance called *cytoplasm*. It is made of water and tiny particles of nutrients and salts. All of the tiny structures of the cell float in the cytoplasm. All of the activities of the busy cell take place in this ooze.

FIGURE 3 *The cell membrane.* Some particles, like water, are so small that they pass easily through the cell membrane. They can move in or out of the cell. Others need to be controlled by the cell or may be too large to pass through by themselves. These may be carried in or out by special parts of the cell membrane.

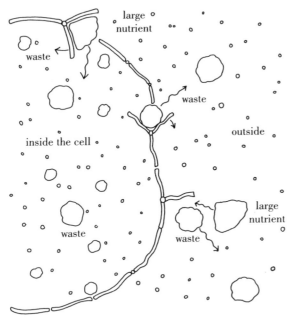

Floating in the center of most plant and animal cells is another membrane-covered mass called the *nucleus*. This is the "control center" for the cell. The instructions, the genes, the blueprints of information, which control growth and development, are held in the nucleus. The nucleus not only controls what the individual cell does, but in organisms made of many cells it works with other nuclei to control how the whole organism grows, changes, and repairs itself.

Scientists are not sure exactly how the nucleus is able to control all of the things a busy cell has to do. Some say it works like a computer. It stores information that the organism will need. The organism doesn't need *all* of this stored information at once. It needs to use information about making repairs only if it is hurt. It needs information about defending itself only if it is being attacked. And the "right time" it finds the piece of information that is needed and gives it to the cell to use.

We do know the nucleus contains instructions for many things the cell must do. It has the information needed for building all the parts of the cell and for making "replacements" when they are needed. It contains information about when the parts should be produced and something about how many should be produced under the right conditions. The control center doesn't just hold the information the cell needs. It must also use the information at the right time. To do that, the nucleus must be able to get information from the rest of the cell and from outside the cell. This information allows the nucleus to recognize when it really is the "right time" to direct the cell to stop old activities or to start new ones. From all the stored pieces of information, the nucleus has to pick the ones that are needed at a particular moment. Then the nucleus has to give that information to the cytoplasm so it can be used. Somehow the nucleus also has to know when that information has done its job and it is time to stop. All of this is being done not by a machine made of metal and plastic and microchips but by the chemical parts of a living organism.

The chemicals that hold the information of the cell—the genes—can sometimes be seen as long tangled threads, called *chromatin*. At certain times these threads coil and tighten into small dark rods called *chromosomes*.

The nucleus of the cell is very powerful. With its storehouse of information and instructions it directs and controls what the cell does. It helps tell the cell what to do and when to do it. It says what to produce and how much of it. It controls what a cell reacts to and how it reacts.

Most of the activity of the cell, however, takes place outside of the nucleus, in the cytoplasm. The cytoplasm is teeming with tiny structures that carry out the many jobs ordered by the nucleus. Most of the structures of the cytoplasm are too small to be seen with regular microscopes. They can be found only with electron microscopes that can magnify objects many thousands of times their normal size.

In this tinier-than-tiny world, there are tubes and tunnels that move materials through the cell. There are storage packets of nutrients, filled with sugars, starches, and fats. There are membrane-covered "sacs" that hold chemicals to help the cell use the energy of stored nutrients, repair itself, or build new parts when needed. There are small, complicated structures that work like little batteries, storing energy for the cell to use when it is needed in a hurry. There are structures so tiny that even with very powerful microscopes they only look like little beads. These are the factories of the cell, the place where the cell is able to manufacture thousands of different protein molecules. These molecules are used by the cell to build its parts and as helpers in hundreds of cell reactions (see figure 4).

The cells in green plants have all these parts plus two more features not found in animal cells—they have a rigid outer surface called a *cell wall,* and they have small green beadlike bodies called *chloroplasts.* The cell wall gives the plant cell its shape and rigidity. The chloroplasts contain the green pigment that gives the plant its color and traps the energy of sunlight so the plant can make its own food.

Where do all these cells come from? According

FIGURE 4 All cells have similar basic structures.

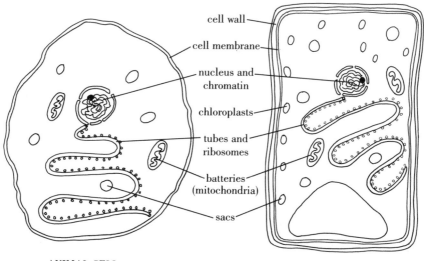

cell wall
cell membrane
nucleus and chromatin
chloroplasts
tubes and ribosomes
batteries (mitochondria)
sacs

ANIMAL CELL

PLANT CELL

to the cell theory, all cells come from other cells. To make more cells, an old cell must be able to make a copy of itself. This is happening in your body all of the time. It happens every time you grow taller. It happens every time you get a cut or a scrape and your body repairs the injury. The growth or repair happens because the cells of your body are dividing to produce more cells. This kind of division takes place trillions of times in your lifetime. Anything that happens that often in nature has to follow a pretty strict set of rules to avoid making mistakes. In the next chapter we'll discuss cell division and the rules that control it.

CHAPTER 3

CELL DIVISION —

MANY FROM

ONE

Have you ever taken a look at pictures of yourself when you were little? You might see that some of your most striking features were there even then—the shape of your eyes, an expression on your face, your dimpled smile. If you have a series of pictures of yourself as you were growing up, you can follow your progress. First you were a helpless little baby; then you were crawling around. Then you took your first shaky steps, and later you became an independent young child. If you look at these pictures carefully you will notice that growing up isn't just a matter of growing bigger. As you grew, you changed.

Your size changes, of course, and so does your shape. Compared to older children, babies seem to have large heads for their bodies. They aren't very well coordinated. They depend on the people around them to take care of their needs. When people say you "sprout" in late childhood, it's because your arms and legs seem to grow so much faster than the rest of you. Even the ways that you think and act change as you grow up. The list of things you can do for yourself now is growing even faster than your limbs. By the time you become an adult, your body will have changed even more. Some of those changes will be outside changes and easy to recognize. Others will take place on the inside, changing your body's food needs, letting you develop larger muscles, and giving you the ability to produce children of your own.

Think for a moment about the tallest, largest, most enormous human being you have ever seen. That person was once a small child and, before that, a tiny baby. How did a baby grow so big?

It becomes even more difficult to imagine all the changes that took place when you realize that each of us started out, not as a baby, but as a single fertilized egg cell. That egg cell was not much different from an ordinary hen's egg—not really too much to look at. Of course a human egg doesn't have a shell and is much, much tinier. It is so small that you cannot see it without a magnifying lens. As you developed inside your mother's body, that one tiny cell became many cells. By the time you were born your body was made up of billions of cells and those cells were no longer all the same. You had many different types of cells. Some made up your skin, others your muscles, bones, and nerves. The cells also had developed some pretty amazing organization. You had eyes and ears, a nose and a mouth, hands with fingers and feet with toes. Inside you had a brain, a complete skeleton, a heart, lungs, a stomach, kidneys, and so on. You were a well-formed, complex, but *small* human being.

Where did all these billions of cells come from? If we all start out as a single fertilized egg cell, how do we manage to have billions of cells in our bodies at birth and trillions of cells by the time we are full grown? There are really only two possible explanations: either the new cells are added on from the outside or somehow the tiny original cell creates more cells from itself. Finding out which explanation is the correct one is not a simple task. Because human babies grow inside the bodies of their mothers it is not easy to tell how these changes take place. Until recently, we could only look at development in other animals to give us a clue about our own.

An animal that develops outside of the body is much easier to study. One that doesn't have a hard, opaque shell makes the job even easier. Frog eggs have been terrific for this kind of observation. They are large enough to see even without a microscope. A frog lays several hundred eggs at a time, so there are many eggs to watch and compare. Most important, they are transparent, and that lets the observer see inside to watch the developing frog change.

By watching frog eggs develop, scientists have found that new cells are not added from the outside. The original cell can split itself in two. Each of those two cells can divide itself in two again, producing four cells. The four cells divide to make eight, the eight can produce sixteen, and so on and so on. You can imagine this going on enough times until there are hundreds and then thousands of cells. The first divisions take place without much rest between them, so each division produces smaller and smaller cells from that original "large" egg cell. After a while, there is a pause between divisions so that cells can grow larger before they split in two. Food stored in and around the egg is used for this kind of growth. In the tadpole that finally bursts out of the egg sac, each cell comes from that original single fertilized egg cell, after a series of divisions and redivisions have taken place.

Very recently scientists have learned to photograph the first hours of development of fertilized egg cells of humans. They have found that we are not so different from frogs! The very tiny human egg cell divides rapidly to produce two cells, then four, then eight, then sixteen, and so on until there are too many to count. The food for growth comes from the mother's body, not from the egg itself. By the time the baby is born, cells have divided many, many times to produce a complete but tiny human!

After the baby is born, the divisions continue. As the body is making new cells, the child grows bigger and taller. Even after most of the growth has slowed down, cell divisions continue. When you scrape your knee or cut your finger, the healing is a result of more cell division. Skin cells near the wound divide to mend the break. Some cells in your body have to be replaced all the time. The red cells in your blood that carry oxygen, don't last very long. They have to be replaced by new red cells every three to four months. The new cells come from the division of special cells in the marrow of your long bones. Over the years you probably replace most of your body cells with new ones. The new ones always come from the division of older cells.

How does nature manage this without making a lot of mistakes? It took scientists a lot of careful study to find out.

Division is an active process carried out by living cells. You might think that the best way to study how cells divide would be to just watch the process happen. That has been done with some success by looking at tiny one-celled organisms, like amoebas. Many of these irregularly shaped creatures can normally be found in stagnant pond water. By adding nutrients and light to a sample of pond water it is possible to end up with hundreds of amoebas in a drop (see figure 5).

FIGURE 5   Amoebas reproduce by splitting in two.

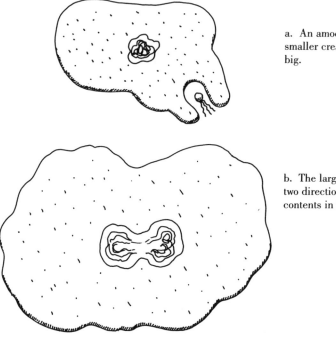

a. An amoeba feeds on smaller creatures and grows big.

b. The large amoeba pulls in two directions, dividing its contents in two equal parts.

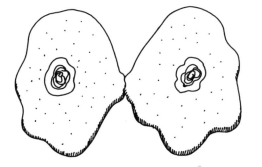

c. The cell membrane pinches between the halves, and two identical amoebas are produced.

They are lots of fun to watch under the microscope. Amoebas move by letting their gooey cytoplasm stream in one direction, stretching the cell membrane to produce a little "foot" projection. The rest of the amoeba just oozes behind this foot. If the amoeba needs to change directions, it just streams off a new lead "foot." A microscopist with patience can find some that are reproducing. Amoebas do this by just splitting in two! First the organism grows bigger than it usually is. Then the amoeba looks as though it is moving in two opposite directions at once. As it stretches, the things inside the amoeba divide equally between the two parts. Then the amoeba uses the cell membrane to pinch off in the center, producing two complete, but slightly smaller than usual, new organisms. These can then start hunting for food so they can grow to normal size.

Observations of living cells let us know that there is some organization to the splitting process. These studies, however, do not give scientists enough information. Looking at a living organism presents several problems. For example, if you watch a living process, you can't stop it to see something more clearly. It doesn't wait for you to see what is happening; it just goes on happening, sometimes pretty quickly. If you think you missed something, you have to find the process happening again and start all over. There is no instant replay. Also, the original studies of this type were done a long time ago. The pictures we have of those original observations are not photos or films; they are drawings and sketches made by the scientists themselves. Sometimes they drew the pictures as they watched what was happening. At other times they drew their pictures later from memory. Another problem with studying living material is that it is often hard to see. It may move around constantly and be very hard to follow or keep in focus, or it may be so thick that light doesn't pass through it well; if so, it may look like a shadow under the microscope. Very few living things have any color when you look at them so closely. Most of what is seen under the microscope is in black and white or shades of gray and silver.

In order to overcome these difficulties, people sometimes look at organisms and parts of organisms in a different way. Instead of looking at living material they look at what is called *fixed* material. Fixed material is no longer alive. It is often preserved with chemicals that allow it to last almost forever.

It is possible to do a lot of things with fixed materials that you could never do with a living specimen without completely destroying it. You can slice off super-thin pieces of fixed materials so you can look at a single layer of cells at a time. You can mix the materials with stains and pigments that give parts of the cell bright colors and make invisible parts of the cell easy to see. Fixed materials can be examined as if they were still photographs of an instant in a living process. You can look at the still photo for hours on end and study all the tiniest details of it.

Of course this way you only see *one* instant in the life of that fixed cell. If you look around carefully, you will find other cells caught at another instant when they were fixed. By putting together a series of these "still photos," cells trapped at different stages in a living process, you can recreate the moving, active, living process, but in much more detail because there is time to study each step very carefully.

When scientists wanted to study the process of cell division, they looked at places where growth was taking place in the organism. In plants, for example, long, tangled roots reach out under the ground in all directions searching for water and minerals. The plants need these raw materials to produce their food, but they cannot move toward them the way an animal would. To reach their nutrients they must *grow* toward them. The root tips of plants are growing almost all the time. When scientists looked at the cells in the growing root tips of plants, they made some startling discoveries.

They saw in the growing area of the root tips many cells just like the ones they were used to seeing in most plants. They were cells, with cell walls, cell membranes, and dark grainy nuclei. There were also lots of cells that looked different. Some of these cells looked large and swollen, but the biggest change was in the nucleus of the cell. In these "different" cells, the membrane surrounding the nucleus was gone. Instead of thick, grainy material in the nucleus, there were threads, or strands. Some cells had long, thin strands, while others had shorter, thicker strands. There were cells that covered the whole range between these two extremes. There also seemed to be a shift in the position of these darkly stained little rods that the scientists named *chromosomes* (see figure 6).

From the "still photos" of these many fixed cells, scientists recreated a "moving" picture of how a single cell divides. The cell swells as it gets ready to divide. The membrane around the nucleus disappears. The material that was inside the nucleus draws together into threads. These threads get shorter and thicker and separate so they are no longer all tangled together. Each chromosome now looks like two threads stuck together at one spot. It is this step that gives cell division its scientific name—*mitosis*. It comes from the Greek language and means "thready condition." These doubled

FIGURE 6 Cell division produces visible changes in the cell.

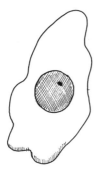  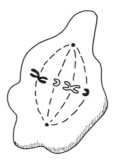

The "resting" cell.

Threadlike chromosomes appear, and the nucleus "dissolves."

Chromosomes line up across the center of the cell.

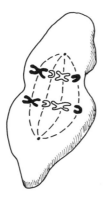 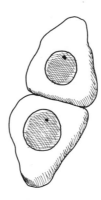

The chromosomes are pulled apart to opposite ends of the cell.

The cell membrane pinches off in the middle, and two new "resting" nuclei appear, one in each new cell.

threads move around until they are lined up near one another in the center of the cell. Each double-stranded chromosome now gets pulled in two directions at once. The "tug-of-war" splits each chromosome apart at its point of attachment, so that the two threads of each chromosome end up at opposite ends of the cell.

Here the rods loosen up again. They are once again surrounded by a nuclear membrane, and the threads quickly disappear. The two nuclei now look like the old grainy nucleus. During the last steps, the rest of the cell becomes very active. The contents stream away from the center, usually distributing themselves pretty equally between the two halves. The cell membrane pinches off between the two nuclei and creates two cells where there used to be one (see activity box 2).

This description doesn't just apply to cells in the growing tip of roots in green plants. It has been seen over and over again in plant and animal cells from many different sources. When cells have to make more cells, this is the way they do it. Is this story just a guess? After all, the story was created from a lot of "still photos" of cells. What if the pictures have been put in the wrong order? Wouldn't the story be different?

Recently it has been possible to check out this idea by watching and filming living cells going through the process of division. New microscopes can use light to highlight the "invisible" parts of cells. This way the parts can be seen without using stains that poison and kill the cell. Cells have been taken out of the organism in which they normally grow and put in bottles where they grow in a thin layer on the side of the glass. These cells divide very rapidly. With the help of the

TWO

UNDERSTANDING CELL DIVISION

Here is a simple activity to help make cell division, or mitosis, easier to understand. Let your fingers be your guides!

YOU WILL NEED   your hands

WHAT TO DO   Hold your folded hands in front of you. They represent the nucleus of the cell just before division begins (picture 1).

Unfold your hands and hold them with the palms flat against each other. The fingers now represent the chromosomes, which copied themselves during the resting stage (picture 2).

Move your hands slowly away from each other with the fingers of your right hand slightly bent toward those of your left hand. Your fingers now represent the dividing chromosomes being separated from each other (picture 3).

Now clench both hands into fists. They now represent the two new identical nuclei (picture 4).

(Adapted from R. West, "Making Mitosis Handy—Helpful Hints," *Science and Children*, Sept. 1987.)

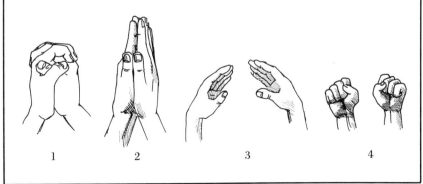

1   2   3   4

special lighting, they have been filmed in the process of dividing. The descriptions of division made by the earlier scientists were right.

The very neat and precise "dance" of division makes sure that each of the two cells, called *daughters,* gets its "fair share" of what was in the original cell. This is especially important, because the nucleus is the control center of the cell. Each new cell gets a copy of everything that was in the nucleus. Each also gets half of the cytoplasm of the swollen cell. This gives them the machinery to carry out the orders of the nucleus. When division is complete they are a little smaller than a regular cell, but they can use nutrients from their surroundings to grow quickly.

Cell division is a living process used mostly for growth and repair. For some organisms, like the amoeba, it is also the way of producing a new generation of amoebas. The "parent" amoeba simply divides in two. This isn't a process that would work well for an elephant.

What *does* happen when the organism is more complicated? What happens when it is made of many cells and has many parts? How are new organisms produced? It is not always as easy as dividing in two, as we will see in the next chapter.

CHAPTER 4

THE NEXT
GENERATION

Nothing lives forever. The closest any creatures have come to immortality are the giant sequoia trees that grow on the West Coast of the United States. These trees can live for thousands of years. Scientists believe that one of them, in a national park in California, is about four thousand years old. That makes it the oldest living organism we have ever found. If it really is this old, it was a sapling when the pyramids were being built in Egypt. Although even these trees will die someday, no other creatures we know of even come close to their life span.

Since they can't live forever, at some point in the life cycle of every species energy must be spent in producing offspring. If there is no next generation, the species will disappear as soon as its present members die.

Reproduction is a very important task, and nature has found many different ways for offspring to be produced. Many small creatures, like the amoebas described earlier, produce a new generation by just dividing in half. The parent grows big, distributes its contents evenly, and pinches in two at the center. The result is two identical organisms where there used to be one. It is impossible to tell the two apart, and neither one can really be called the "parent." There are just two offspring.

The parent doesn't have to disappear for offspring to be produced by some kind of splitting process. In some organisms, like the microscopic hydra, the baby grows right out of the side of the parent. The miniature hydra, called a *bud*, feeds off of the parent until it has grown large enough to have a good chance of surviving on its own. When it is ready, it just breaks away from the parent's side. Then it floats off, to settle somewhere nearby and start feeding on its own. When it has grown large enough, it, too, can form a bud and produce its own offspring (see figure 7).

Scientists recently discovered a small desert lizard that produces its offspring in a way that is very unusual for large, complex organisms. A single cell from the parent is enough to start a new lizard. This egglike cell goes through all the changes that were described in the last chapter for frogs' eggs, even though it has not been fertilized by a second parent. By rapid division, the one cell becomes hundreds and then thousands of cells. The hollow ball of cells folds in upon itself forming layers. By moving, sliding, and changing into different cell types, this ball of cells eventually forms itself into all the parts of the lizard.

In all of the cases described so far, the offspring have only one parent. Their traits come only from that one individual. In fact, each offspring in these cases is an exact copy of its parent. There are only two ways that they can turn

FIGURE 7   An adult hydra with a bud growing out of its side. When the bud is big enough to survive on its own, it breaks off and floats to a new spot. The hydra is too small to see well without the aid of magnification.

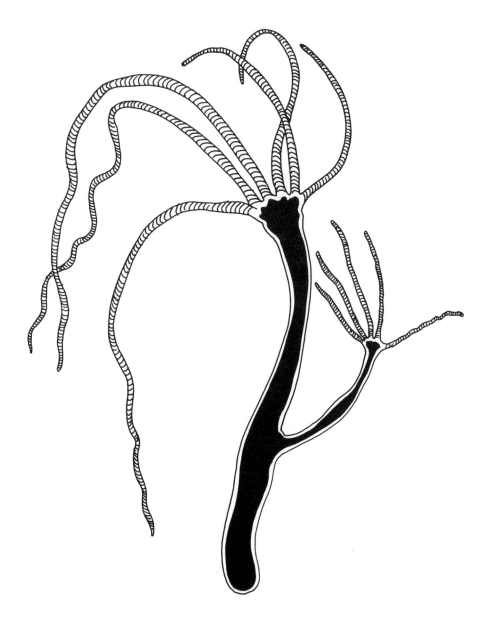

out different: (1) the instructions they get from their parent could change by accident; (2) something serious could happen in their environment to alter how they need to react. Although both of these things can happen, they occur only rarely. Most of the time these organisms develop exactly the way their parent developed. They are as much alike as identical twins.

This is not true for animals and plants that have two parents, however. If their two parents are not identical to each other, the offspring are not exactly like either one. These organisms share in a wider *variety* of traits than those that have only a single parent. Each parent has the same kinds of general features, like eyes and ears, nose and mouth, but the forms of the trait—their exact appearance—may be quite different. The offspring do not get *all* the specific traits possessed by one parent. Instead, it appears that they get some traits from each parent. Sometimes it seems the forms of the trait in the two parents blend in the offspring.

The careful study of how traits are passed from two parents to their offspring began about 130 years ago. An Austrian monk named Gregor Mendel combined his work as a gardener in the monastery with his curiosity as a scientist about inheritance. The organism he studied was the pea plant. While working in the garden he noticed that peas have many different traits. All pea plants have stems and leaves, flowers, pea pods, and peas. That doesn't mean that all pea plants are the same. Some pea plants are tall (6 feet) and some are short (1 foot). Some have green pods and some have yellow pods. Some have smooth seeds and some have wrinkled seeds. In all, he found seven traits of his pea plants that could appear in two very different ways (see figure 8). Usually his pea plants produced more offspring like themselves. Tall pea plants produced more

FIGURE 8 After several years of research, Mendel settled on seven characteristics that displayed clearly contrasting forms.

TRAIT	ALTERNATE FORMS		
SEED SHAPE	smooth	wrinkled	Seeds were smooth and round *or* angular and wrinkled.
COTYLEDON COLOR	yellow	green	Seed leaves were yellow *or* green.
SEED-COAT COLOR	white	gray	Seed coat was white *or* gray.
POD SHAPE	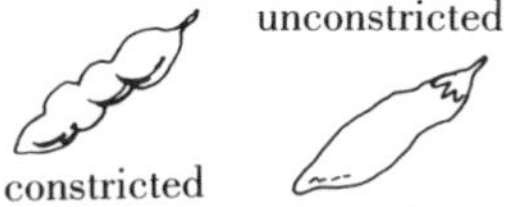 constricted	unconstricted	Ripe pods were constricted *or* not constricted.
POD COLOR	 yellow	green	Unripe pods were yellow *or* green.
POD POSITION	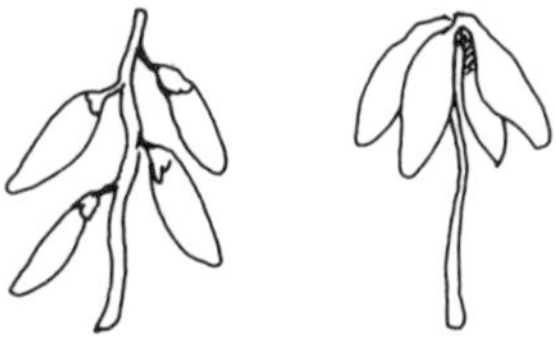 along stem	top of stem	Flowers and pods were located all along the stem *or* only at the top of the stem.
STEM LENGTH	tall	short	Pea plants were tall *or* short.

tall plants. Short pea plants produced more short plants. Mendel called any plant that only produced more plants like itself "pure" for that form of the trait. A tall pea plant that came from tall parents and produced only tall offspring was considered a "pure" tall pea plant (see figure 9). Mendel was curious to see what would happen if he produced offspring from parents that were each "pure" for a different form of the same trait. What would the offspring look like?

To satisfy his curiosity Mendel designed some very simple experiments. He changed the way his pea plants normally produced seeds so that *he* could be in control. He removed the anthers, or pollen producers, from the flowers of his test plants. When he wanted to fertilize these flowers he brushed pollen onto their pistils from the plant *he* selected to be the other parent. The seeds he produced could then be collected and planted to see what kind of pea plant *they* would produce.

When Mendel produced offspring from parents pure for two different forms of the same trait, he called the offspring *hybrids*, just as we do today. What did Mendel find out by studying these hybrids? First he found out that the traits of the parents did not blend in the hybrids. This was a surprise, because it was what most people believed at the time. A six-foot-tall pea plant crossed with a one-foot-tall pea plant did *not* produce a three- or four-foot-tall pea plant. The hybrids for the tallness trait were all six feet tall. Nor did a yellow-podded pea plant crossed with a green-podded pea plant produce a pale-green-podded pea plant. The hybrids for the trait of pod color always had green pods. One form of the trait was always "stronger" than the other. He called this form *dominant*. All of the hybrid offspring showed that form of the trait. The form of

FIGURE 9 A plant was considered "pure" for a trait if all its relatives showed the same trait.

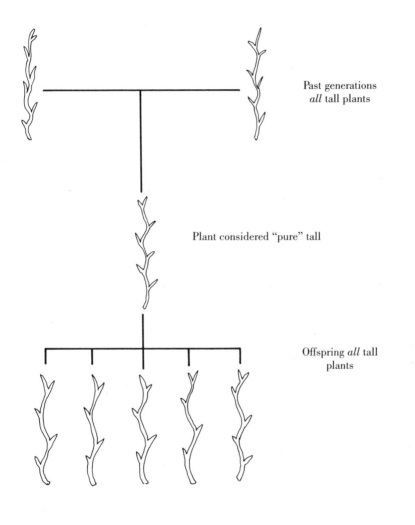

Past generations *all* tall plants

Plant considered "pure" tall

Offspring *all* tall plants

the trait that did not show up was called *recessive*. At first Mendel wasn't sure if the "weaker," recessive form was hidden or had disappeared entirely. To find out, he used two hybrid plants as the parents of a new generation. In each experiment, the hidden form of the trait reappeared in *some* of the offspring of the next generation (see figure 10).

FIGURE 10   Crossing two pure plants with opposite forms of a trait produces identical hybrids. Offspring of the hybrids produce *some* offspring that have the "hidden" recessive form.

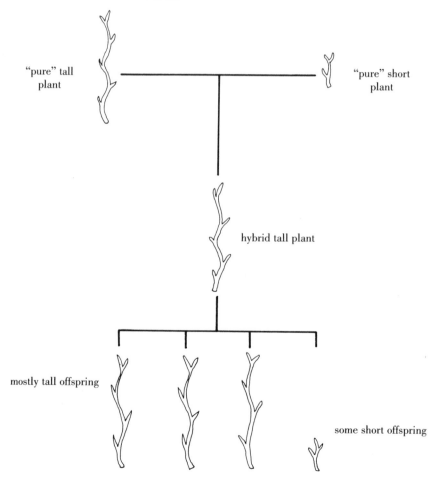

"pure" tall plant

"pure" short plant

hybrid tall plant

mostly tall offspring

some short offspring

He repeated these experiments over and over again to be sure his results were not an accident. He tried the experiments reversing the plant that gave the pollen and the one that gave the flower. The results were the same. He repeated the experiments with each of the seven traits he had selected. Later he repeated the experiments using more than one experimental trait at a time (see figure 11).

After all of his experiments were finished, Mendel had to try to explain all of the things he had discovered. He knew that certain things were passing from the parents to their offspring, and he called those things "factors." Mendel said that the factors controlled how the traits developed in the offspring. We now call these factors *genes*. Mendel argued that each organism has a pair of genes that controls the development of a specific trait. When they produce an offspring, each parent gives it *one* gene from each pair it has. In the offspring, the two genes (one from each parent) act together. If the two genes are not of the same type, one gene may be dominant. That means it is stronger in controlling the development of the offspring, and it determines the form of the trait that will show in the hybrid. The hybrid looks like the parent with the dominant form of the trait. He said the hidden gene will reappear in future generations if it is combined with another recessive gene like itself.

He also found that all seven of the traits he looked at were inherited independently from one another. Offspring from a six-foot-tall green-podded pea plant crossed with a one-foot-tall yellow-podded pea plant would all be six-foot-tall green-podded pea plants, because these are the dominant forms of the traits. But in the next generation, the offspring of these hybrid plants could be six feet tall and green podded;

FIGURE 11 Results of the hybrid cross.

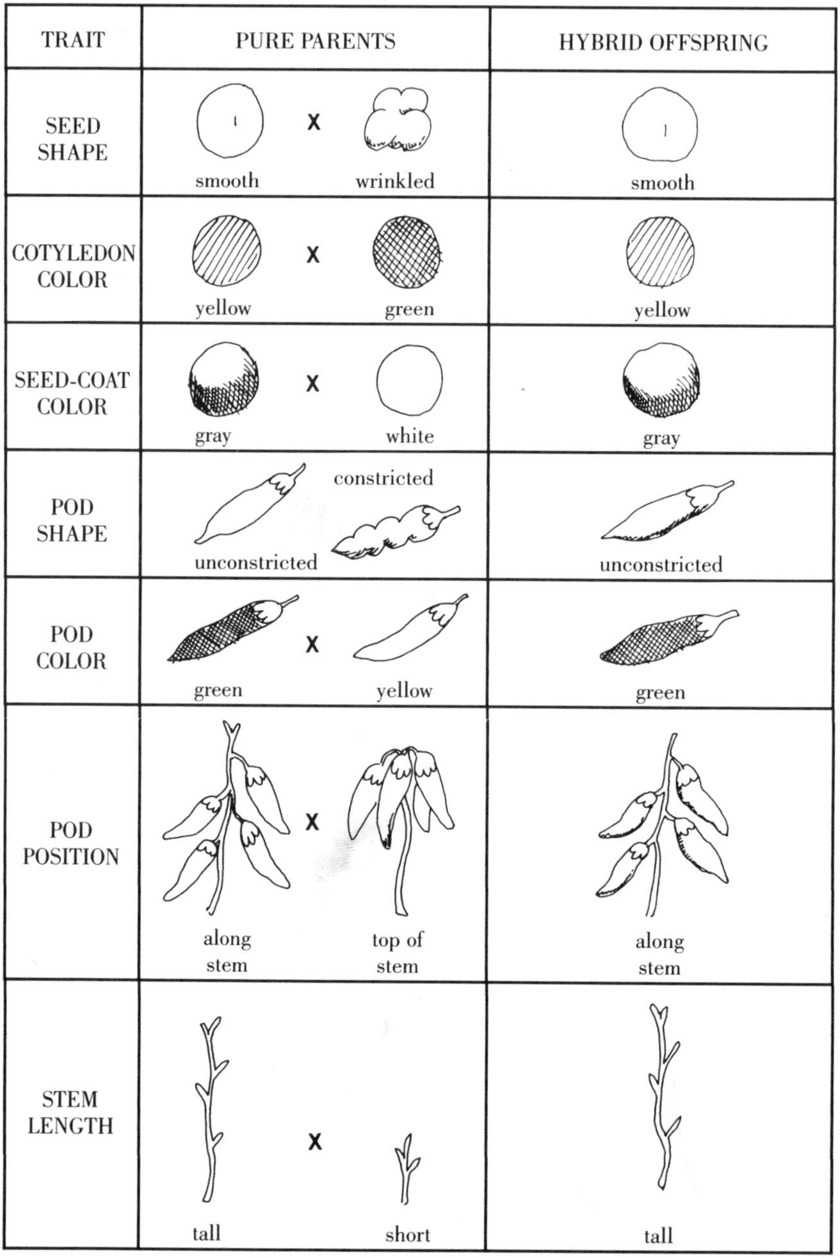

TRAIT	PURE PARENTS	HYBRID OFFSPRING
SEED SHAPE	smooth X wrinkled	smooth
COTYLEDON COLOR	yellow X green	yellow
SEED-COAT COLOR	gray X white	gray
POD SHAPE	constricted / unconstricted	unconstricted
POD COLOR	green X yellow	green
POD POSITION	along stem X top of stem	along stem
STEM LENGTH	tall X short	tall

six feet tall and yellow podded; one foot tall and green podded; or one foot tall and yellow podded. Traits could be mixed up and recombined in later generations (see figure 12).

As time passed, inheritance was studied in more animals and plants. Scientists wanted to know if Mendel's ideas explained all kinds of inheritance. It became clear that sometimes Mendel's simple picture of how inheritance worked was not enough. New experiments showed that his ideas didn't explain all situations.

Sometimes there is no dominant gene, and a hybrid form really is a blend of the traits of both parents. A red-flowered morning glory crossed with a white-flowered morning glory will produce a pink-flowered plant, just as though you were mixing paints. Unlike mixed paints, however, the original red and white traits can reappear in the next generation of flowers if the pink hybrids are crossed.

Sometimes traits do not occur in simple contrasting forms. Although pea plants may be *either* one foot tall *or* six feet tall, there are few other organisms where height appears in such extremes. There are very tall humans and very short humans, but these aren't the only two possibilities. Human height is a continuous range from very short to very tall and everything in between.

It also became clear that the genes don't work on their own. Just because an organism receives tallness genes from its parents doesn't mean it will be tall. A pea plant seed with genes that say it can grow to be six feet tall may not reach its full height if it is forced to grow in poor soil with little water or not enough sunlight. On the other hand, no matter how good the environment, a seed with genes to grow one foot tall can't stretch much beyond that limit. Genes give instructions for

FIGURE 12 Genes are inherited independently in Mendel's pea plants. Traits that appear together in individuals of one generation may be separated and recombined in later generations.

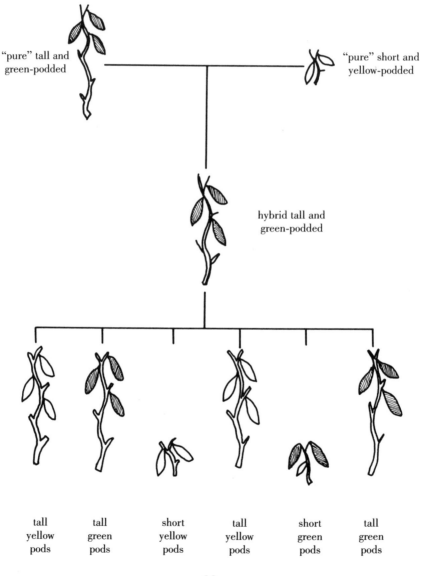

what *can* be done, not necessarily what will be sure to happen.

We now know that Mendel's great work was a simplification. He assumed that each trait is controlled by instructions in a single pair of genes. It has been hard to get people to give up this idea. Geneticists now understand that very few traits are really controlled this way. Most traits, in fact, are influenced by many pairs of genes acting together. All genes also have to work with the environment in which the organism grows. How tall *you* will grow depends on a dozen or more genes that act together. Since you got those genes from your parents, you can use their height as a clue about how tall you could grow. Since you mix and combine their genes, you can end up looking quite different from them. Short parents can produce taller children and tall parents can produce shorter children. How well you eat is going to influence your height as well.

Even though we have had to modify some of his ideas, Mendel did lead the way into modern genetics. He said the evidence from inheritance showed that all genes came in pairs. This helped people see the importance of paired chromosomes in inheritance. Then he said that these gene pairs regularly split up. This happens as the parent's body gets ready to pass its information on to offspring through cells known as *germ* cells. If this is true, then there has to be another kind of division besides regular mitosis, which we discussed in Chapter 3. Somehow, as the germ cells are formed, there has to be a way to cut the number of genes in half. In theory this reduction makes a lot of sense. It explains why only part of the information from each parent seems to be passed along to each offspring. It can help people to understand why an offspring can have one parental trait and not have another. It also explains why one offspring can get a trait from a parent while another offspring of the same parent does not.

Supppose both parents did pass on all of their information to their children. Each parent would give the children two genes for each trait. That would mean all their children would have four genes for each trait. In the next generation the children would get four genes from each parent and so would have eight copies in each of their cells. In the next generation there would be sixteen of each gene, then thirty-two, and so on. The nucleus would be a very crowded place. This way doesn't match what we observe. Each generation maintains the same total number of chromosomes and the same number of genes as its parents had.

Scientists have looked at the cells that are passed on to the next generation in many plants and animals. They have found there is a special kind of cell division to form these cells. It is different from the cell division for growth and repair that was described in Chapter 3. By adding an extra division, the germ cells reduce their number of chromosomes, which are the carriers of genes, to half. Since it cuts, or reduces, the number of chromosomes, this division is called *reduction division,* or *meiosis* (from the Greek word meaning "to make smaller"). It is exactly what Mendel said was needed to explain the rules of inheritance he observed (see figure 13).

Notice the simple changes in the steps of the process that allow reduction to take place. The chromosomes start out looking the same as they would look in a regular cell division. They are made of two strands connected at a single point. The main difference is that two divisions, instead of one, take place to separate these chromosomes. In the first step, the chromosomes line up in the center of the cell, opposite an

FIGURE 13 Reduction division, or meiosis, has two separations of the chromosomes. This produces four cells, each with half of the number of chromosomes found in the original cells.

The cell gets ready to divide. Doubled chromosome rods appear.

The chromosomes line up in the center of the cell *with* a partner chromosome.

The first division separates the partners into different cells.

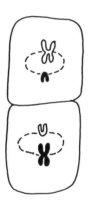

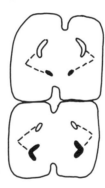

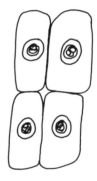

The chromosomes line up in the center of the two cells just as in regular cell division (mitosis).

The threads of the chromosomes are pulled apart into two different cells.

Four cells are produced. Each one has half the number of chromosomes of the original cell.

identical chromosome partner. This division pulls the partners apart into different cells. Then a second division occurs in each of these two cells. This one is just like a regular mitotic division. The chromosomes line up in single file in the center of the cell. This division separates the two strands of each chromosome into two different cells. The result is four cells, each of which contains half the number of chromosomes and, therefore, half the genes of the original cell.

This meiotic division doesn't occur as often as regular mitotic division. It is only used to produce those very specialized reproductive cells known as germ cells, or *gametes*. Each parent produces gametes that contain *half* of its genetic information. In sexual reproduction, the gametes of the two parents are brought together and merge with each other. This merger produces a fertilized egg or a seed. In the nucleus of this cell, the two half-sets of chromosomes, one from each parent, combine to make a new whole set. The number of genes in the offspring cell is now the same as that of the parents. The combined cell contains a complete set of instructions for the organism, but it is a new mixture, partly from each parent. The fertilized egg, through a long series of mitotic divisions, can go on to develop into a new individual, similar to both parents but not identical to either.

It is different when you begin to study human inheritance. It is *not* like studying pea plants. No one controls the selection of parents or the number of children produced. No one sets up experiments to "see what would happen if . . ." To study human genetics, scientists must use families that happen by chance. They collect as much information as they can about families and the traits they possess. There is a way to record all the information known about a family so it is easy to understand and easy to study. It is known as a family tree, and you will learn how to draw one in the next chapter.

CHAPTER 5

IN THE SHADE

OF YOUR

FAMILY TREE

Genetics is the study of genes, the bits of information and instructions that we all get from our parents. It is the study of how these genes can direct and control our growth and development. But you cannot see a gene. What you can see are traits, characteristics, how people look, and how they develop. This is called their *phenotype*. The actual color of your eyes—black, brown, hazel, green, blue, or gray—is *your* phenotype for the trait of eye color.

Although appearance, or phenotype, is usually related to genes, it is not exactly the same thing. Genes work with one another and with the environment to produce a final appearance. It is not always possible to know what kind of genes have been at work just by looking. Short people may be short because they have genes for shortness or because they don't have many genes for tallness or because their environment didn't let them reach their full height.

Even with simpler organisms than humans it

isn't so easy. If you took one of Mendel's six-foot-tall pea plants, you wouldn't know just by looking at it what its genes were like. It has two genes that control its height. It must have at least *one* tallness gene. This would be enough to produce a six-foot-tall plant. But the tall plant could also have *two* genes for tallness.

If you really wanted to know about the plant's second gene, there are two ways you could find out about it. Both ways involve finding something out about the plant's "family." One looks back at the past generations of plants. The other looks forward to the offspring of the plant.

First, the parents of the tall plant could give you a clue to its genes. You could look at Mendel's scientific records. If one of its parents had been a short plant, you would know that your tall plant was a hybrid. Its second gene for height would have to be a short gene. The short parent had *only* short genes to give to its offspring. Even though it is hidden in the tall appearance of the hybrid, that gene *must* be there. The gene pair, or *genotype*, of this plant is one tall (T) and one short (t) gene.

The second way is to look at the plant's off-spring. If any of its seeds produce short plants, you would know it had a hidden short gene to pass on.

When it comes to studying human inheritance, the job can be much more difficult. Sometimes the simple methods used by Mendel can still be helpful. With people, there are certainly lots of traits that could be studied. The list of characteristics that are found in humans is enormous. But not all the traits that are found in people are inherited from their families. People can change their appearance in many ways that may have nothing to do with their genes.

FIGURE 14 Looking at an organism's "family" to help identify its genotype.

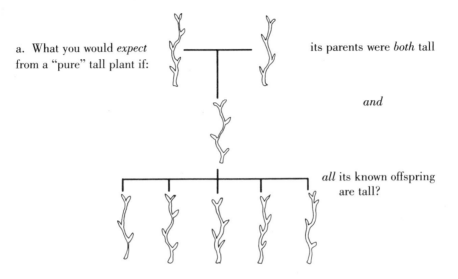

a. What you would *expect* from a "pure" tall plant if:

its parents were *both* tall

and

all its known offspring are tall?

b. What would you hope to find to identify a hybrid if:

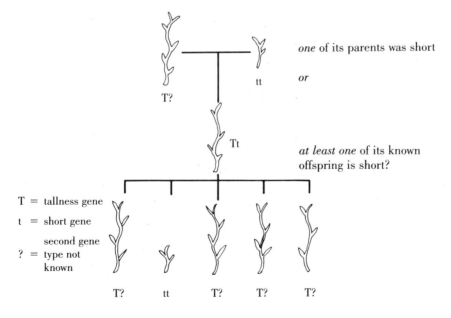

one of its parents was short

or

at least one of its known offspring is short?

T = tallness gene
t = short gene
? = second gene type not known

There are people, for example, who have pierced ears or pierced noses or tattoos on their bodies. Sometimes several members of a family will choose the same way of decorating themselves. A number of relatives may end up with the same appearance, but does this make it a genetic or inherited trait? There is no evidence that any of these are. They are added on, by choice, after birth.

The same is certainly true about a person's hair color. We know this trait is under genetic control, but it can be dramatically changed. A person may start out with black hair and dye it blond, or start with blond hair and tint it red. When finding out about family traits, you have to be careful to find out what is "natural" and what has been altered.

Many traits in humans are partly genetic and partly the result of the environment in which the person grows and develops. A person's weight and height are partly controlled by the action of a number of gene pairs working together. Both of these traits are also greatly affected by the person's eating habits and nutrition. These and many other traits do not have sharp or distinct forms but rather come in gentle gradations within a range of possibilities.

Most of what is known about human inheritance comes from interviewing people about their family traits. It has been possible to find out who has certain traits in a family. By studying many families with the same trait scientists can discover some things about how that trait is passed along from generation to generation. These family studies have taught us a lot about ourselves. To keep records of these interviews, scientists draw a pedigree, or family tree, whenever they investigate a family trait.

This is how a pedigree is drawn. Every person

in the family is represented by a symbol, usually with that person's name written just below it. All females (girls and women) are shown as a circle: ○ . All males (boys and men) are shown as a square: ☐ . The symbols are connected to one another by lines that show how they are related. A man and a woman who are married or who have had children together are connected by a straight line between their symbols: ☐——○ . All brothers and sisters, or *siblings*, are connected by a line that runs above their symbols: ⌐ ┬ ┬ ⌐ . The parents are connected to their children by a straight up-and-down line between the "marriage" line and the "sibling" line: ⊤ . The family relationships can be understood without having to write a lot of words on the drawing.

Generations are kept separate from one another by placing all brothers and sisters (and their partners) on the same line. All of their children (the next generation) form a line below their parents. Grandparents are on a line above their own children. Generations form straight lines across the page, with the most recent generation being at the bottom of the page and the most ancient generation that has been traced at the top of the page.

This is how the pedigree, or family tree, of an imaginary family might be drawn.

Sarah is the imaginary subject of our interview. We first ask her about her immediate family. She has a sister and a mother and father. When the family tree is drawn, the symbol for Sarah is always marked with an arrow. This part of her family tree would look like this: ☐⊤○
○ ○
↑

We can add on to this family tree one branch at a time. Sarah's mother has a brother who is married and has two children, a boy and a girl. When they are added on, Sarah's family tree looks like this:

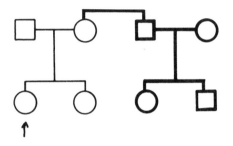

Sarah's father has an unmarried brother and a sister who is married with three children. The sister has two girls and a boy. When they are added on, Sarah's family tree looks like this:

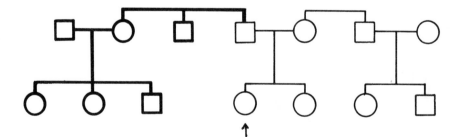

Now we can add both sets of grandparents, and the tree grows upward to this:

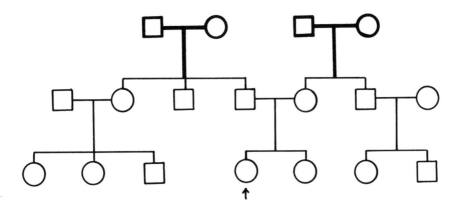

Sarah happens to know a lot about her family. She is able to add information about her grandparents' brothers and sisters (her great-aunts and great-uncles) and their families. She can also tell the interviewer a little bit about her great-grandparents. When the family tree is finished, it looks like figure 15.

All of the relatives she told the interviewer about are included in her family tree. They have been given numbers instead of names to make it easier to read the chart. These are the relatives Sarah has in her family tree:

        siblings: sister (7)
        parents: mother (17), father (16)
        aunts: (14, 19)
        uncles: (13, 15, 18)
        first cousins: girls (3, 4, 8); boys (5, 9)
        grandparents: grandmothers (24, 26);
                          grandfathers (23, 25)

great-grandparents: great-grandmothers
(29, 31, 33, 35);
great-grandfathers
(28, 30, 32, 34)

second cousins: (1, 2)

The pedigree is finished when all the individuals and relationships you want to include have been drawn in. Now you can start putting in the record of the trait you are studying. A description of the form of the trait, or phenotype, of each person in the family tree is written at his or her symbol. How you get this information depends a lot on the family and on the trait being traced. Sometimes the trait is something that anyone can see, like eye color or a cleft chin. In a case like this, the trait may be traced by just asking people what they know about themselves and about other members of the family.

Sometimes the trait is more complicated. If you are studying blood groups, each person's phenotype will have to be determined by a laboratory test. If you are studying the inheritance of a disease with very specific symptoms, you might need an examination by someone trained in its diagnosis. In this case each individual in the family has to be seen and tested individually. Sometimes this can be complicated to arrange when people live far away from one another. This is one of the things that makes human genetics so difficult.

The family tree is used to keep a record of who in the family has what form of the trait. You can list more than one trait on the same pedigree if you want. The idea is to be clear. If it gets confusing because there is too much information on the chart, it is better to make two or more charts. The form of the trait each person has may be written near his or her symbol in words. It may also be represented by dots, or shading,

FIGURE 15 Sarah's completed family tree or pedigree.

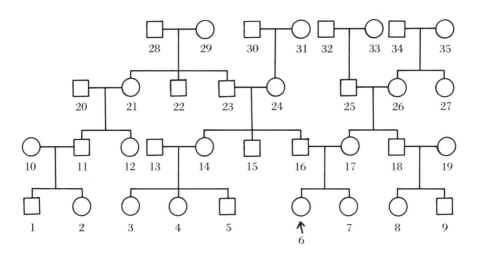

or letters inside the circles and squares. Each pedigree should include a key so anyone reading it can understand it.

The family tree of a particular trait can be used by scientists for two purposes. First, it can be used with many other similar family trees to try to decide *if* a trait is genetically controlled at all. If it seems to be genetic, the pedigree can help decide how it is inherited. This is where a lot of information about human inheritance comes from (see figure 16).

If the trait being studied is *known* to be genetic, the family tree can be used in another way—to help figure out what genes a particular family member is carrying. If the pattern of inheritance is already understood, you can make a best guess about the genotype of family members by looking at the phenotypes of their nearest relatives, as we did with Mendel's pea plant at the beginning of this chapter (see figure 17).

Let's take a look at the inheritance of two

FIGURE 16   A group of pedigrees can be used to figure out the method of inheritance of a human trait.

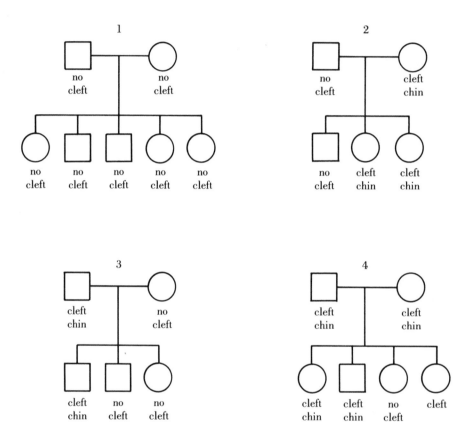

Geneticists carefully study the pattern of how traits appear in families to decide what kinds of genes control the trait. If both parents have no cleft in their chin, they produce *only* children with no cleft chin (as in family 1). This family is "pure" for the trait. When both parents have cleft chins (as in family 4), they sometimes produce a child with no cleft chin. The parents are passing on a *hidden* trait to their offspring. If this pattern is seen again and again in many families, it would seem the gene for cleft chin trait is dominant (C) and the gene for no cleft chin is recessive (c).

FIGURE 17 If the inheritance pattern of a genetic trait is already known, the family tree can be used to assign genotypes to family members.

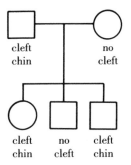

The mother with no cleft and the son with no cleft must have two recessive genes for no cleft (cc).

Since the father produced a son with no cleft, he must have one hidden recessive gene to pass to this son. Since he has a cleft chin he must also have one dominant gene for the cleft (C). His genotype must be Cc.

The daughter and the other son must have one dominant gene (C), because they both have cleft chins. They got this gene from their father. Their mother only has recessive genes to give to her children. These two children must have Cc genotypes, too.

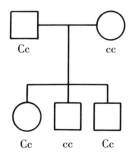

genetic traits in our imaginary family, one for blood group and the other for eye color. With this information about the relatives added to it, Sarah's family tree looks like figure 18.

Many different blood groups can be identified in people. We are looking at the ABO-group in Sarah's family. This is the first blood group that was discovered. The ABO trait is controlled by a single pair of genes in each individual. The gene itself can take three different forms: an A-gene, a B-gene, and an O-gene. Depending on what two genes a person has, the phenotype of that person's blood can be one of four types—A, B, AB, or O. The blood type is identified by a special laboratory

FIGURE 18   Sarah's family tree with blood types and eye color traits added for each family member.

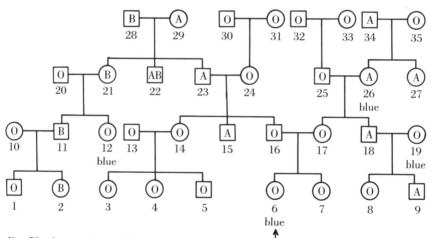

Key: Blood groups in symbols A, B, O, AB. Eye color below symbol for blue eyes. All others have dark eyes.

test. Blood typing is very important when a person needs a blood transfusion. People in hospitals got sick after being given the wrong type of blood. This was how blood groups were originally discovered.

The four different blood types are produced by combining the three types of genes of the ABO-group that people can get from their parents. A person who has only O-genes will have O blood. The O-gene is the most common gene around, but it is recessive and will be hidden any time it is paired with an A- or a B-gene. A person with one A-gene (whether the other gene is A or O) will have A blood. A person with one B-gene (whether the other gene is B or O) will have B blood. A person with one A-gene and one B-gene will have AB blood. The A- and the B-gene are called *codominant* because neither one will hide the other (see figure 19).

FIGURE 19   People have A, B, AB, or O type blood. Which type they have depends on the gene they got from each of their parents.

Gene from mother

		A	B	O
	A	AA A-type	BA AB-type	OA A-type
Gene from father	B	AB AB-type	BB B-type	OB B-type
	O	AO A-type	BO B-type	OO O-type

The A- and the B-gene can be "seen" in every person who has those genes. This makes them easy to trace as they are passed along in a family. Let's see how these two genes have been inherited in Sarah's family tree.

Great-grandpa (28) had a B-gene that he passed on to one daughter (21) and one son (22). His daughter passed this B-gene to her son (11), who passed it on to his daughter (6). By chance, the B-gene never got to Sarah's part of the family tree (see figure 20).

There were two sources of A-genes in Sarah's family. Great-grandma (29) had an A-gene that she passed on to both of her sons (22 and 23). Her son (23) passed it on to his son (15). Great-grandma (34) also had an A-gene, which she passed on to both of her daughters (26 and 27). One of these daughters passed the gene on to her son (18) and he passed it along to his son (9). Again by chance, neither of these genes reached Sarah's immediate family.

Inheritance of the gene for eye color is not as well understood or as easy to trace as that of the ABO blood group. A single gene pair may be responsible for eyes being dark pigmented (black, brown, or hazel) or light pigmented eyes (blue, green, or gray). But acting with that pair may be several other genes, most of them responsible for skin color, which make small changes, giving us dozens of different eye colors.

Look back at the pedigree of our imaginary family. Most people have brown or dark-pigmented eyes. Just by looking at these people, we cannot tell if they have one gene for dark eyes or two. Only two members of Sarah's immediate family have light eyes—Sarah (6) and her maternal grandmother (her mother's mother), (26). Her uncle's wife (19) also has blue

FIGURE 20

a. Tracing the B-gene through part of Sarah's family tree

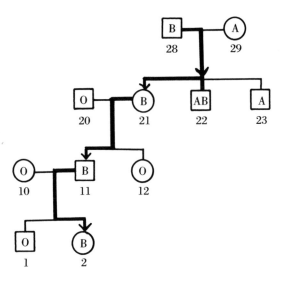

b. Tracing the A-gene through part of Sarah's family tree

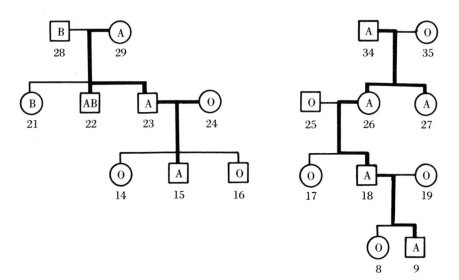

eyes, but she is not directly related to Sarah and cannot have passed her genes on to her.

We can tell more about the genes for eye color in Sarah's family if we study the pedigree carefully. Since the gene for light eyes is a recessive gene, Sarah must have two of these genes. She had to get one from each parent. Even though both of her parents have dark eyes we now know they must be hybrids for this trait. Her mother (17) must have received her hidden gene from *her* mother, who had no other kind of gene to give her children. It is not possible to be sure where the light-eyed gene from Sarah's father (16) came from. It *may* have come from *his* father. He has a cousin from that branch of the family who also has blue eyes. This would only be a guess.

Try making your own family tree and follow a trait or two through its branches (see activity box 3, 4).

What are these genes, that they have so much power over our development? Think of all the things they control. We have discussed their control over traits like hair color and eye color. If you think about it, these traits are not really the important ones. They are not basic to our survival.

Without having to be taught how, our bodies can digest food, process nutrients, carry oxygen from our lungs, and control our inside temperature. Many more things that *are* crucial to our survival are also part of what we inherit from our parents. These processes are so numerous that we could probably never learn to do them on purpose. How some of this control works is the topic of the next chapter.

BUILDING YOUR OWN FAMILY TREE

Recording how traits are inherited begins with drawing a family tree. With a little patience, and answers from members of your family, you can build your family tree.

YOU WILL NEED some paper with lines (graph paper works best)
a pencil with an eraser
a ruler

WHAT TO DO Begin the tree with your immediate family and work your way up and out from there. Use a pencil so you can erase mistakes. Remember you will probably draw the pedigree at least twice. The first time you just want to get all the people and relationships down on paper. The second time you can spread them out more evenly, placing them so large families aren't all squished together.

Place the paper sideways, with the long edges at the top and bottom. Start by drawing the symbol for yourself (a square if you are a boy, a circle if you are a girl) just about in the center of the page and about two inches from the bottom edge. Using the right connections for relationships, add your brothers and sisters and write their names under their symbols. Draw a line up from your sibling line and add your mother and father. If either of your parents has or had another marriage, add your stepparent on the level with your parent. Then add your half brothers and half sisters on the level with your own symbol. On your parents' line, add the symbols for the brothers and sisters of your mother and father (your aunts and uncles). Add your parents' parents on the level above them. You can now add the husbands and wives of your aunts and uncles and, down on your level, your first cousins.

boys and men

girls and women

brothers and sisters (siblings)

parents

line to children

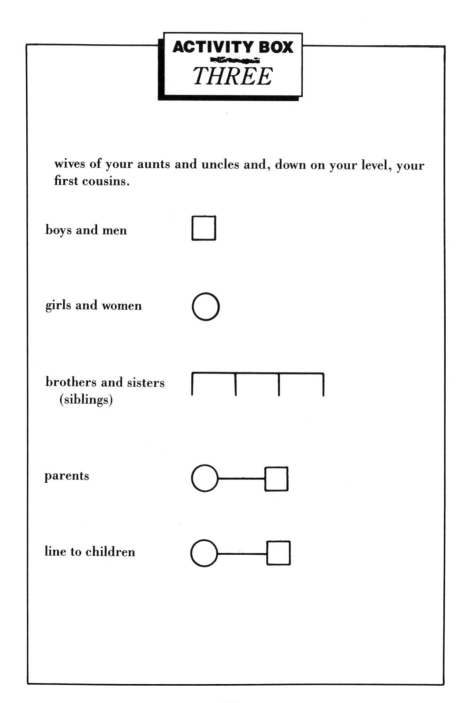

ACTIVITY BOX
FOUR

TRACING HOW TRAITS HAVE BEEN PASSED ALONG IN YOUR FAMILY TREE

YOU WILL NEED a copy of your family tree drawn in activity box 3
a pencil with an eraser

WHAT TO DO Look at the chart below. It lists human traits that are at least partly under genetic control. Make sure that you understand what the two alternate forms look like before you try to trace their inheritance in your family. Once you are sure you understand the trait or traits you have chosen, you may begin by interviewing the members of your family about their features. If members of your family are not available, you can use another relative's best memory of that person's trait. Record on your family tree the trait or traits of each member.

Dominant trait	Recessive trait
A. B, or AB blood group	O blood group
Cleft chin	No cleft chin
Dimples	No dimples
Dark-Pigmented eyes (brown, black, hazel)	Light-pigmented eyes (blue, green, grey)
Dark hair	Blond hair or Red hair

CHAPTER 6

HUNTING FOR

THE BLUEPRINT

Staying alive isn't an easy job for any organism, even the simplest ones. There is so much that must be done. Somehow every organism must collect energy from the environment. Some can do this by trapping energy directly from the sun and making their own food. If they can't do that, they still have to eat. That means hunting for their food. Whatever they trap (energy, raw materials, or food) usually can't be used right away. It has to be changed or digested. There is energy to be stored for later use, and there is waste to be dumped.

Not everything organisms meet in their environment is good for them. There are bad things, like poisons, that they must avoid. Organisms can get sick or injured. Then there will be body parts that need repair or replacement.

While one organism is busy hunting for its food, there may be another organism around who would like to use *it* as food. To avoid being eaten, organisms may need a way to recognize their hunter. Then they may try to get away or hide. If they can't do that, they might protect themselves with sharp thorns, a tough shell, or something that makes them taste bad.

All organisms die eventually. Therefore, if the species is to survive, organisms must produce more of their own kind (see figure 21).

This is a lot for an organism to "know" how to do. How is it possible to "learn" to do all of these things quickly enough? Fortunately, each individual organism doesn't have to figure it all out on its own. The solutions to some of these problems of life are built into the organism. The inherited instructions that direct their development already hold some of the answers on how to survive.

What are these instructions, or "blueprints," like? How would anyone begin to look for them? Individual cells and whole organisms need so many different kinds of information. How is all of it stored? How does the cell find the piece of information it needs? If the information was written out in the form of a book, it would probably be gigantic. But all of that information must somehow fit into these very tiny cells that make up all living things.

In addition to being small, the "blueprint" has to be easy to copy. Every time a cell divides, each daughter cell needs to get its own complete set. This means that before a cell divides it must be able to reproduce the blueprint exactly. During division, the two sets must separate from each other carefully. Each daughter cell must get everything it needs— nothing extra, nothing missing. That way, what was "known" by the original cell is "known" by both of the new cells.

This gave people something to look for. Nobody

FIGURE 21   *The jobs of living things.* Even the tiniest of living organisms have many tasks to perform in order to survive. Their bodies are organized to carry out each job successfully.

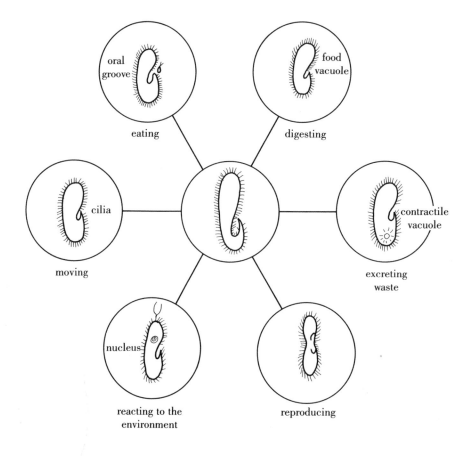

knew what the blueprint itself looked like. Now they knew something it *had* to be able to do. That was a good place to start the search. Of course, they started with the parts of the cell they already knew about. Now they looked at them more carefully, trying to figure out where the blueprint could be hidden.

What did they know about this mysterious blueprint? They knew it was small. It had to be the same in every cell of an individual. It was probably pretty much the same in all organisms of the same species. From Mendel, they also knew that, as in many organisms, it was probably "packaged" in pairs, one from each parent. During cell division whatever held the blueprint would have to be doubled and then divided evenly between the daughter cells. The blueprint would probably have to be "packaged" in some way so its parts couldn't get lost or torn.

The most logical candidates for the job of carrying the blueprint were the chromosomes. They were the only part of the cell that met all of the requirements.

The patterned, careful separation of the chromosomes during cell division, which we discussed earlier, was one of the things that made them seem suitable as carriers. This separation process would explain how the message could be doubled and divided evenly, and the number of mistakes could be kept low.

If people were right and the chromosomes did carry the cell's blueprint for life, there were still a lot of questions to be answered. Where is the message to be found in these tiny rods? How is it stored? How is it used? Figuring out the answers to these questions took a lot of people a long time. They are still looking for some of the answers today.

The first thing they tried to find out was which part of the chromosome actually held the information. If they could get an idea of what the information looked like they could make some better guesses about how it worked. Billions of cells were broken open in the laboratory and the chromosomes inside of them were collected in tubes. Special machines analyzed what kinds of chemicals the chromosomes were made of. The machines broke the chromosomes into small pieces and separated them. They found several different kinds of protein in the chromosomes. They also found a very long molecule, with a very long scientific name. They called it DNA for short. They looked at chromosomes from different organisms. They found different amounts of protein and different types of protein. But that DNA molecule was always there. It was so consistent that the search for the blueprint focused there.

How could scientists be sure this DNA molecule was the real carrier of the genetic blueprint? How could it hold such a complicated set of instructions? Could this molecule make perfect copies of itself, something the blueprint had to be able to do? How could information held in a molecule be used by a living cell to control all of the cell's activities?

At first there was no way at all to even imagine the answers to these questions. No one had any idea what a DNA molecule looked like. "Looked" is really the wrong word to use anyway. DNA is a small part of a small part of a tiny cell that can itself barely be seen with a microscope. How do you examine, analyze, or diagram something that you can't see or feel?

It was very difficult to solve the problem of DNA's structure and behavior. It took a lot of careful detective work. Many different techniques, old ones and new ones, had to be used to find parts of the answer. Some of the techniques told scientists about the smaller pieces that made up the DNA molecule—what kind of pieces they have, how many of each kind. Other techniques told them things about the shape and size of DNA. Other techniques gave hints about how the parts of the molecule fit into each other. Putting all of these bits and pieces of information together was like working on a jigsaw puzzle. It was made harder because no one had any idea what the final picture should look like. It took brilliant minds to put the puzzle pieces together so they made sense.

After many years of trying to fit all the evidence together, scientists finally came up with a clear idea of what DNA is like. Since no one has actually seen the molecule, our idea of it is called a *model*. Scientists discovered that it looks a bit like a spiral staircase with a handrail (see figure 22).

The spiral staircase of DNA (deoxyribonucleic acid) turns out to be made up of a very interesting combination of chemicals. The "handrail" of the molecule is made of a chain of very simple sugar molecules linked by a chemical called a phosphate. Sticking out from each of the handrails are pieces of the "stairs." These pieces are called *bases*. To make a stair you need two bases, one from each handrail. Four different kinds of bases make up the staircase. These are usually called by the first letters of their names: A for adenine; C for cytosine; G for guanine; and T for thymine. It turns out that a stair is not made of just *any* two bases stuck together. Only two kinds of pairs work to make a stair: A and T or C and G. It seems that the bases are shaped in such a way that they will not link tightly to each other and hold the staircase in place unless they fit together correctly. A fits with T and C fits with G in just the right way.

FIGURE 22 *The "spiral staircase" of DNA.* The bases form the "stairs" of the spiral. They fit together tightly and link only when A is opposite T, and C is opposite G. The "handrail" of the spiral is a chain of sugar and phosphate.

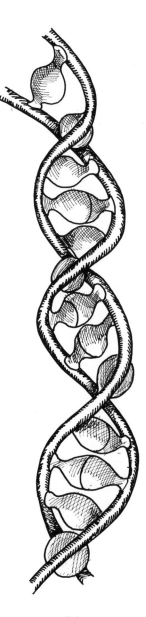

This is a very interesting structure, but where are the instructions for putting it together? If DNA is really the master of the cell, telling it what to do and how and when to do it, where does it get this power? The key appears to be in the bases that make up the staircase. The information needed by the living organism is stored in the order of the four bases, A, C, G, and T. Think of the bases as letters of the alphabet. The twenty-six letters of our alphabet can be put together in different order and in different combinations to create words, sentences, and books that have different meanings. In the same way the four bases of DNA can be put together in different order and in different lengths to hold all the information that a cell or the organism needs. A chain of bases, which are linked together in a specific order and which work together, forms a gene. Many genes linked together in a chain are what we see under the microscope as a chromosome. In different organisms, the basic molecule that holds the information is usually DNA. The information that is *coded* in the order of the bases makes the genes different. This code determines what kind of organism develops.

How does the cell make an exact copy of a long complicated chain of DNA? It has to do this before it divides so that all daughter cells will have a complete set of instructions after separation occurs. The spiral staircase of DNA actually copies itself (see figure 23). When it needs to double, it unzips down the center, splitting the steps at the link between the bases. Once the chain is opened, the bases on either side are without partners. We already know the bases fit together like pieces of a jigsaw puzzle. Each exposed base can hook up to *only* one other base: A with T, C with G, G with C, and T with A. Once the spiral has opened, each "half staircase" helps line

FIGURE 23   DNA building blocks fit together like pieces of a toy, A with T and C with G. When the chain of DNA must be copied, the two parts of the chain separate at the base pairs. Free bases floating around in the cell line up with their bases on the opened chain partner and create two identical chains from one.

up the free bases that are floating around in the cell. An exposed A grabs on to a free T, an exposed C grabs on to a free G, and so on. Each half finds free bases to replace its missing side. The cell then uses special chemicals to hook these assembled free bases together. The "handrail" is then connected by linking the sugars and phosphates. Now two identical staircases (each one half "old" and half "new") have been formed from the one original DNA chain. These two chains, packaged in the linked strands of a chromosome, are pulled apart in cell division. This way they end up in different daughter cells.

Except for a brief time during cell division, when the nuclear membrane disappears, the long molecule of DNA is always inside the nucleus of the cell. Most of the cell's activity takes place in the cytoplasm. How does trapped DNA control the development of that cell and the organism as a whole? We will explore this question in the next chapter.

CHAPTER 7

THE BUSY

RESTING CELL

When you watch the process of cell division you see lots of activity. Chromosomes appear out of the grains of the nucleus and move about the cell. Cytoplasm streams. Even the cell membrane gets involved, moving, pinching, building a new barrier between the two forming cells.

When a cell is not dividing, things look so much calmer. The nucleus just looks grainy, not busy. The cytoplasm still streams around, but it doesn't look so purposeful; it is just circulating. Things look pretty peaceful. That's probably why this period between divisions has been called the "resting stage" in the cell's life cycle. The term paints a pretty clear picture. The process of cell division is hard work for a cell, and between divisions the cell has to rest to regain its strength for the next division. Sort of like runners resting between races to get their wind back. But this is really not an accurate picture of what is happening in the cell. In fact, it really gets the picture all backwards.

Cell division is not the "purpose" for the existence of most cells. It is a necessary *part* of what they must do, but it is only a part. Cells usually divide because growth is still taking place in the organism. Or there may be an injury that needs repairing. Or there is a need to replace old cells with new, rejuvenated cells. Cells spend the rest of their time doing a lot of other things.

If it is part of a larger organism, what the cell does will depend on what kind of cell it is and where it is located. It may be a blood cell that carries oxygen or defends the organism from infections. It may be a nerve cell that sends messages to and from different parts of the body or stores memories. It could be a cell in the stomach that produces acids and enzymes to digest food, or a cell in the intestine that absorbs the tiny digested nutrients. It could be a special stinger cell that helps catch prey. It could be a cell that produces a bad taste or a poison that makes the organism unfit to eat. Cells have to do all of the many tasks of living things that have been described throughout this book. It may look peaceful from the outside, but the "resting stage" of the cell is anything but restful!

We have already said that the instructions for carrying out all of this activity are held in the chemical known as DNA. These long strands of DNA hold a lot of information for the cell to use. Unless the cell can get that information out of the nucleus and into the cytoplasm, it is not very helpful. Once in the cytoplasm the information has to be translated into action by the cell. We said the "fit-together" nature of the bases is the key to copying DNA. It also enables this molecule to control the activity of the cell.

By unzipping a part of the DNA spiral known

as a gene, the cell can take out some of the information stored there. It is as though the cell borrows one book from the library of information stored in the DNA. But the gene itself can't be removed from the long chain of DNA and it can't leave the nucleus. Instead, the needed part of the DNA chain is copied into another molecule called mRNA. This can move from the nucleus into the cytoplasm (see figure 24).

The chain of mRNA is very similar to DNA. There are four main differences: (1) mRNA is shorter; (2) it is only one strand instead of two spiraled together; (3) mRNA has a different sugar; and (4) one of the four bases is different, in that U substitutes in mRNA for all the T's in DNA. Any of these differences may be what lets mRNA leave the nucleus while DNA is trapped there. The chain of mRNA doesn't last as long as DNA. It is destroyed by the cell after a short time in the cytoplasm. While they are there, however, the copies of genes in mRNA are used in the cytoplasm to control the activity of the cell. How do they do this?

A lot of what an organism does involves molecules known as proteins. There are thousands of different kinds of protein molecules inside you. Proteins make up the membranes of your cells. There are proteins in all the tiny structures inside your cells. Proteins help to digest food. They speed up and slow down chemical reactions in the body. They help in the construction of new building materials. They identify invaders and help defend you from infection. They carry important messages from one part of your body to another.

In your body, proteins control your eating and sleeping patterns. They control how you grow up. They control your body's reaction to illness and how you age. They are

FIGURE 24 *Making a gene.* When the cell needs to use the information stored in a gene, that part of the DNA opens. The gene is copied into a strand of mRNA. This molecule carries the gene message to the cytoplasm.

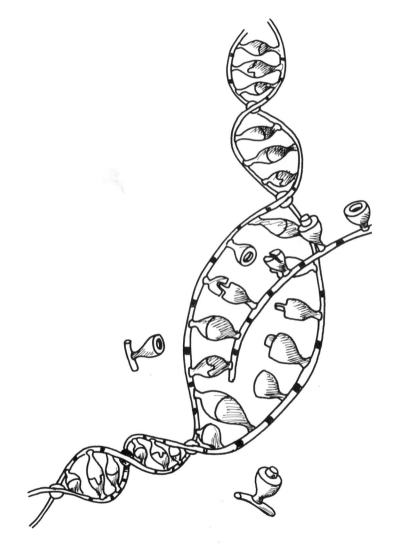

certainly amazing molecules. All of these different proteins are manufactured in the cytoplasm of your cells.

Since proteins are so powerful, whatever controls them is powerful, too. This is what the DNA molecule is doing. It controls which protein is produced, when it is produced, and how much of it is produced. To understand how DNA manages this, we have to understand a little bit about what proteins are like.

All living things are made of building blocks called cells. Proteins are also made of small building blocks. These are called amino acids. There are twenty amino acids found in nature. They are all made of atoms of carbon, hydrogen, oxygen, and nitrogen, but each of the twenty is a little different from the others. All proteins are chains of these amino acids linked together. Some proteins are small, made of only a few amino acids. Some are huge, made of hundreds of amino acids linked together. Which amino acids are in the chain and what order they are in determine what a protein is and what it can do.

There are always free amino acids floating around in the cell. By selecting the right amino acids in the right order, the cell can construct any protein chain it needs. A cell may use the pool of free amino acids to produce the protein fibrin, which forms a scab when you cut yourself. With different directions a cell can use the same pool to produce a protein called growth hormone, which orders your bones to grow longer. Cells use the same basic building blocks, in a different pattern, to produce useful proteins that do entirely different jobs. It is the *order* of the amino acids that is so important.

This is where the DNA comes in. The order of

the bases A, C, G, and T in the DNA molecule is a code for the order of the amino acids in the protein the cell needs to make. To build the protein, the cell has to "read" the DNA code. This is how it works. The gene is copied from the DNA into a chain of mRNA. Now *it* has the coded message for the protein. The mRNA is the gene's messenger to the cell factory. It leaves the nucleus and enters the cytoplasm (see figure 25).

FIGURE 25 *Making the protein.* The gene messenger, mRNA, finds its way to a ribosome in the cytoplasm. This is the factory of the cell. The ribosome "reads" the message and lines up tRNAs and the amino acids they pull along with them. The cell quickly joins the amino acids to each other, and the protein is complete.

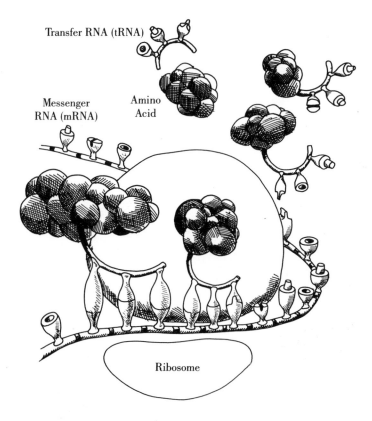

Transfer RNA (tRNA)

Messenger RNA (mRNA)

Amino Acid

Ribosome

Each of the twenty amino acids has its special code in the mRNA. But there is still a problem to be overcome: the amino acids can't understand the message by themselves. They need helpers to guide them to the right spot. These helpers work like translators. They are actually small pieces of tRNA. One end hooks on to specific amino acids. The other end is able to recognize part of the code in the messenger chain of mRNA. When the code calls for *their* amino acid, the tRNA moves into place, dragging its attached amino acid with it. Each tRNA lines up in the right order with other tRNAs. An enzyme in the cell quickly attaches the lined-up amino acids to one another and releases them from the tRNAs.

The protein chain grows longer as the tRNAs drag the right amino acids over and line them up in order. When the mRNA has been "read" from one end to the other, the protein is done and can be used by the cell.

The DNA is in control, but not directly. *It* controls the sequence of bases in its messenger, mRNA. The mRNA is understood by the tRNAs that line up in order along its length. The tRNAs drag along the right amino acids, which build the coded protein (see figure 26).

There is a lot of copying and assembling going on in the organism. The fact that so many copies of molecules are made cuts down on the effect of mistakes. The A fits with T and the C fits with G so well that the unzipped DNA molecule can reproduce itself accurately. When a gene is copied into mRNA, the C fits with G, the A fits with T on the DNA, and *U* fits with A on the DNA, and again the copy is accurate. Finally, each tRNA matches its part of the mRNA strand, so that the amino acids line up in the right order.

If everything follows the pattern correctly,

FIGURE 26   DNA is not directly involved in the production of proteins. DNA produces a copy of a gene in the molecule mRNA. It acts like a messenger, carrying the information of the gene to the ribosomes where proteins are produced. This happens when tRNA drags amino acids into the right order.

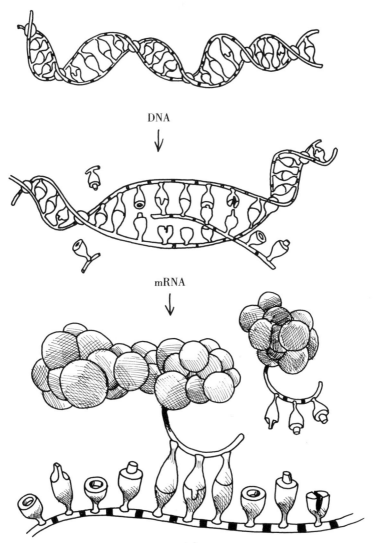

DNA

mRNA

protein

there are no mistakes. But we know nothing can be counted on *never* to make mistakes. The copying and reading of DNA take place more times than anyone can count. There are a lot of chances to make mistakes. Each time, cell division means copying DNA. That happens many, many times as you go from a single fertilized egg cell to an adult with trillions of cells.

We can't even begin to guess how many times the genes in your DNA are copied into mRNA. Every cell in your body needs to use some of the information stored in your genes. Some cells use more of this information than others; some use it more often than others. Somewhere between 50,000 and 100,000 genes are stored in human chromosomes. Most of them probably get used by some cell at some time.

The last step, building the protein itself, isn't mistake-proof either. A mistaken reading by tRNA can put the wrong amino acid in the chain.

So how do things turn out okay if so many mistakes could be made so easily? The reason seems to be that not every mistake is really important. There may be lots of opportunity for mistakes, but most of them don't make any difference.

What kind of mistake *could* cause a problem? The troublemaker would probably be a bad protein—a protein that could not do its job. There are three places where this kind of mistake could occur: (1) in building the protein; (2) in the messenger from the nucleus; or (3) in the gene itself. How does the body keep these mistakes from overwhelming it? It seems there are a lot of safeguards built into the cell. These make it harder for a mistake to be really harmful.

Some of the safeguards are built right into proteins. Only a small part of each protein molecule is needed to do its real job in the cell. Part of the molecule is there to give the protein shape. The rest of the molecule works almost like packing material to protect the active, important parts. It doesn't absorb physical shocks, but it can "absorb" the chemical changes that might harm the active parts of the protein.

The organism also protects itself by producing large numbers of molecules of the same type of protein. If a few of these are produced with mistakes in them, they probably won't even be noticed by the organism. The proteins aren't kept in the body forever, either. They are replaced by new molecules of the same type. If the turnover is rapid, a bad protein won't be there long before it is destroyed.

It is also possible for a mistake to occur in the messenger that comes from the nucleus. If this happens, all of the molecules produced from that messenger could have mistakes in them. Here, too, there are safeguards for the organism. Some changes in the messenger *do not* change the message. There is usually more than one way for the messenger to code for the same amino acid, so a small change can be tolerated. There is also safety in numbers among the messengers. The organism produces many copies of the messenger. If one of them is wrong, its impact can be lost in the many that are correct. The messages are also discarded quickly by the cell and replaced by new ones if needed.

The DNA molecule itself can also have mistakes in it. After all, the molecule is copied millions of times in your lifetime. There are safeguards here as well. Whether the mistake here is harmful depends on *where* the mistake occurs on the DNA chain. Much of the DNA chain doesn't actually code for proteins. We don't know exactly what all of this DNA does. It may have something to say about when genes start to work. It may turn them off when their work is done. It may do things we haven't even imagined yet. The point is that not all of the DNA is actively producing proteins. There are long stretches where a mistake in the DNA chain might not be "noticed" by the organism at all.

It also matters *when* the mistake was made in the DNA. This will decide which cell or cells in the body have the mistake. Only the cells produced *after* the original error was made would copy it into their DNA chains. The later the mistake occurs in the development of the organism, the better the chances for normal growth. That way very few cells contain the mistake. In general, the fewer cells that have the wrong DNA message, the smaller its impact.

If there is a mistake in a section of DNA in a group of cells that don't normally use the information stored there, then again the mistake won't matter. If a gene that codes for a digestive enzyme produced by liver cells is wrong in the DNA of some skin cells, why would it matter? The cells with the mistake never need that gene anyway.

There is one situation where changed DNA is very likely to cause problems. That is when the error in the DNA occurs before the organism is conceived. If the change is in one of the cells received from the parents, then the DNA is "wrong" in every cell of the organism's body. If DNA from only one parent is changed, the organism has a second "normal" copy of the gene to work with—the copy that came from the other parent. Sometimes this is enough to allow the organism to develop normally. In some really crucial genes there's more than one copy of the gene in the chromosomes from *each* parent. Then if there is a problem in one of the copies of the gene, there are other "good" copies to fall back on.

Even with all of these safeguards, however, there are still times when something goes wrong. A crucial gene in the development and growth of an organism does not work right. In this case the organism develops improperly. These problems are called genetic diseases. Some of them cause minor problems. Some are so serious that the organism dies. Compared to the number of times that mistakes could occur, organisms with genetic diseases are rare.

The "mistakes" in DNA are not all *bad* for living creatures. Within all of the "normal," healthy organisms there is a lot of variety in the code of DNA that is hidden. The changes are tolerable; they don't cause problems. This hidden variation can come to the surface when conditions change in an environment, though. These changes demand something new from the organisms that live in it. The "something new" may be already there in the changed DNA. Little changes in DNA are the source of variety in organisms on our planet. Over the course of the 3½ billion year history of life on earth, the pressures of a changing environment have pulled from this variation an enormous number of different organisms. This variety is the topic of the next chapter.

CHAPTER 8

FINDING

THE CONNECTIONS

Walk along the edge of a pond and notice the variety of life you can find there. A dozen different kinds of trees may grow nearby. Beneath them, amid the fallen leaves, are many feathery little mosses and a wide variety of mushrooms and toadstools. A spider creeps along its filmy web. Ants and worms crawl over and under the ground. Small mammals—like rabbits, chipmunks, and squirrels—can be seen in the undergrowth. Birds fly overhead and settle onto branches or into the tall grasses that grow around the pond.

The pond itself is teeming with life. Insects and birds swoop down over the pond. Waterfowl swim on its surface. Every so often they dip their heads deep into the water to feed on pond grasses. They are joined on the surface by water-skating insects. These insects stay afloat by not breaking the surface tension of the water. They skim across it hunting for other insects that are trapped there. Below the surface, tadpoles and fish dart through the water. Water spiders move up and down stems of grass, carrying with them a precious bubble of air to breathe. At the bottom of the pond, digging into the mud, leeches and hunting fish wait for their dinner to swim by.

All the members of this pond community seem to fit together neatly. Everything has its special place. Each organism has a slightly different way of finding shelter and food. There are tree dwellers and ground dwellers, creatures of the land and creatures of the water. There are plants that trap the energy of the sun. There are animals that eat plants and animals that eat those animals. There are even organisms, like mold and mushrooms, that feed off of dead plants and animals. Near a pond you will find lots of different kinds of organisms with lots of solutions to the problems of staying alive.

Making sense of all this variety is not an easy task. For centuries naturalists have examined, described, and classified living organisms. They have grouped them according to their characteristics and traits. They have looked for patterns and they have found them. Families of organisms have been found whose members, like human families, look alike and behave alike. The systems of classification have changed over the years. As people examined organisms more carefully or with new instruments they recognized different patterns, discovered new similarities and differences. Even with the changes, a family tree of living creatures can be described. Close branches are organisms that share the most similarities, distant branches have less in common. All of the limbs and branches and twigs of the family tree are made of *living organisms*.

But how do you know if something is alive? Because it eats food, digests it, and gets rid of its wastes?

Because it breathes gases? Because it grows bigger? Because it can repair itself when it is sick or injured? Because it can react to protect itself? Because it can make more creatures like itself? There is *so much* that distinguishes the living from the nonliving world. There are many characteristics you could use. Some of these characteristics are the ways living things behave. There are more similarities than that. You can see some of them with a microscope; other similarities are too small to be seen even with a microscope.

Under the microscope you can see that all living creatures are made of cells. Though there are differences between the outward appearance of cells, when you look closer, you can see that they are all very much alike. Membranes and tiny cellular structures look pretty much the same no matter where they come from. So do the chemicals of inheritance and control, DNA and RNA. Even the biological building blocks— like proteins, fats, sugars, and vitamins—are remarkably alike in all living creatures. With all the differences that might exist, the similarities are stronger.

When you look at the way different organisms develop, you find more similarities. We all start out as a single cell. Even the most complex organism develops after many, many divisions of a single cell. Even *how* a cell divides follows the same pattern throughout the living world. As it develops, each organism grows and takes on the special, characteristic shape and form of its species. No living organism is separate from its environment. We all must get nourishment from the world around us. To survive, each living creature must have a way of sensing the environment and reacting to it. If we reach the age that marks the end of the normal life span of our species, we begin to age and die.

All organisms are a part of a great flow of energy and nutrients. Ultimately we all depend on the energy of the sun to power our activities. Green plants can collect and use that energy directly. They are called the *producers*. The plant-eating animals get their energy "secondhand" and are known as *consumers*. Energy is passed to meat-eating animals from the organisms *they* consume. There are organisms that get their energy from dead plants and animals, and they are known as *decomposers*. What we pass along to one another is not only energy but nutrients. Our planet contains over ninety natural elements, but most things that fill our cells are made of only six of them. The atoms and molecules that make up the chemistry of life are all remarkably similar in the living world.

Everything we have discussed in this book has been about similarities and differences among living creatures. Scientists are convinced that the similarities between living organisms are far more profound than the differences. These similarities tell us that somehow we are all connected to one another. It is not an accident that we are so much alike. Somewhere in the distant past, when life began on earth, the family tree of living organisms began to form. Living things are so much alike because we are all related to one another.

No one was there when life began on earth. What has been recreated of this history comes from trying to understand and explain pieces of information from many sources. Biologists, scientists who study the various creatures living today, provide only part of the information. Geologists, who study the earth's form and its changes, provide some, too. They gather information about the planet's history from the structure of the rocks. Paleontologists, who study fossil records of ancient life forms, provide some more. They can trace the changes and development of the ancestors of the organisms that share the planet today, and they study extinct life forms, organisms that are lost forever (see activity box 5). They work with astronomers and oceanographers and geneticists to get more facts and more ideas. Together they try to solve the great mystery of how life evolved upon this planet.

Scientists now believe that the earth began in a cold dark swirling cloud. The baby sun and planets grew as masses of dust and rock collided and stuck to each other. The cloud of dust may have come from the explosion of a star somewhere else in the galaxy. Eventually the center of our solar system grew big enough and hot enough that it began to glow. A great explosion probably accompanied the sun "turning on." This probably swept away extra dust and even the atmosphere of the smaller inner planets. Things may have heated up for the planet, too. Large meteors kept crashing into it, and radioactive materials decayed inside the young planet, releasing energy.

A period of intense volcanic activity covered the planet's surface with molten lava. The volcanoes also released gases and water that had been trapped inside the planet. These gases created a new atmosphere. Storm clouds and lightning bolts filled the skies of the new planet from time to time. Rain fell and helped cool the hot surface. The water evaporated again quickly and was recycled into clouds and rain. When enough water accumulated and the planet cooled off, oceans and seas formed.

All of this began about 4.6 billion years ago. In the first billion years of the history of the earth, the rock record shows no signs of life. Then the record begins to change. We can only speculate about the earliest forms of life. They may have gobbled up large nutrient molecules directly from the

FIVE

MAKING YOUR OWN FOSSILS

Fossils are records of organisms from the past. They are found in the rocks of the earth. Dead organisms may get trapped in dirt and silt that settles in still bodies of water. The organism leaves an impression in the rock. This may become a space in the rock that can be found millions of years later when the rock is broken open.

YOU WILL NEED plaster of paris
enough water to make a soft paste
foil folded into a shallow box
a leaf, feather, or some other object
    to become your fossil
Vaseline

WHAT TO DO Mix the plaster of paris with water until it is a soft paste. Pour it into the box of foil. Allow it to set for less than two minutes. Smear the bottom of the leaf or feather with a light film of Vaseline. Press the leaf or feather into the plaster of paris firmly, but do not allow it to be covered with the plaster. Allow the plaster of paris to become firm. Remove the leaf or feather. Allow the plaster of paris to dry completely. The impression that is left behind is like the fossils that are found in the rocks.

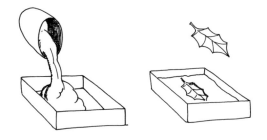

earth's ancient seas. Even in these earliest stages, changes in environmental stress demanded new responses from the life forms living in the environment. If directly available nutrients became scarce, how would the new life forms survive? Organisms that could trap the sun's energy and create nutrients from simpler molecules could survive and prosper.

The rock records do let us know that mats of blue-green algae eventually covered the early oceans. The algae dominated the planet for the next 2.5 to 3 billion years. Their very existence changed the conditions of the planet. Over aeons, other organisms evolved. Some became more complicated. Each new development changed still further the conditions under which new life forms would live. The atmosphere changed, too, becoming more oxygen-rich. This condition eventually supported the evolution of animal forms.

As more living forms were added to the environment, more varieties became possible. Variety supported the development of more variety. Changes in the environment caused changes in the life forms. The life forms also changed the environment. Sometimes the changes were in the physical environment, as when the atmosphere became oxygen-rich. Sometimes the change was in the biological environment, as when the appearance of a new predator sharply decreased the number of prey organisms. Sometimes the changes to the environment were so drastic that many species of organisms could not adjust quickly enough, and they were lost forever; they became extinct.

It is hard to think about what is "quick" in evolutionary time. Since the earth is estimated to be 4.6 billion years old, a short time and a long time in the earth's history isn't like a short time or a long time in your life. The year between your birthdays may seem like a long time, but is less than the blink of an eye in the history of the earth (see figure 27). We say evolution has changed the forms of life that inhabit the earth. We are talking about changes that may take millions or even hundreds of millions of years to accomplish. How do we know how these changes may have happened? They certainly take too long for us to watch them happen. As models for how evolution works, we use what we know about changes we *can* see and even changes we create ourselves.

FIGURE 27   The earth formed 4.6 billion years ago. To understand how the evolution of life fits into this time, imagine the earth formed only an hour ago, at 12:00. On this scale, the first humans appeared in the last quarter of a second!

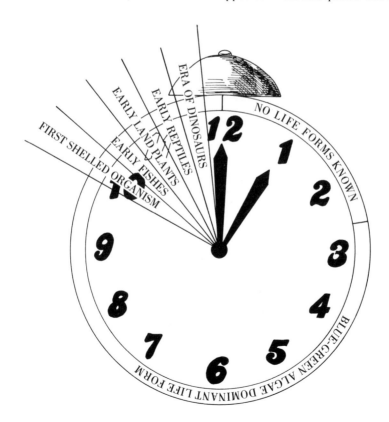

In an earlier chapter, we learned how farmers produce changes in the plants and animals they raise. Carefully picking the parents for the next generation is called *artificial selection*. Scientists have shown that the environment makes similar changes in plants and animals. The changes happen over a longer period of time in a process known as *natural selection*.

The environment doesn't "pick" animals and plants the way a farmer does. The environment doesn't have a plan or even an idea for a specific type of "better" creature. But all organisms must meet the demands of the environment to survive. If an organism has a trait that makes it better able to do this, it might have a greater chance of surviving to produce young. If this happens, the environment is partly screening the parents of the next generation. This is not as fast or efficient as the farmer who can decide absolutely which organisms will be parents and which will not. Over a longer time, the environment can help shape a variety of species whose features make them well suited to survive in it.

We have on rare occasions witnessed real selection taking place. It is an unusual event, because evolutionary changes do not often take place in a human lifetime. However, people have changed the natural environment quickly, especially since the industrial revolution. Sometimes this has produced rapid changes in other organisms in response. Some organisms have been killed off completely by our changes. Others have managed to adapt. The classic example of observed natural selection involves a species of moth in England (see figure 28).

Before the industrialization of England, this moth lived in a light-colored environment. It was hunted by

FIGURE 28 Changing environments cause changing selective pressures. When the environment provided mostly a light-colored background the darker moths were easier to spot than the light moths. The dark ones were gobbled up by hungry birds and often had no opportunity to produce young.

As industrial smoke and soot darkened the background the dark moths suddenly had the advantage of camouflage. Now the lighter ones were more likely to be eaten by the birds and thus prevented from reproducing.

DO IDENTICAL GENES ALWAYS PRODUCE IDENTICAL ORGANISMS?

Genes are not the only things that determine how an organism will turn out. They provide the instructions the organism needs to survive. Yet the instructions can do nothing on their own. They must work in and with the environment to allow survival. How can you tell how important the environment is to an organism's development? You can see how different environments can change the growth of genetically identical individuals.

It is not always so easy to find organisms with identical genes. One place you can find them is on a potato. The "eyes" of the potato can grow into new potato plants. Since the eyes grow out of the same potato—like buds, not seeds—all of their genes are the same.

YOU WILL NEED an old potato with several eyes
a knife
several waxed paper cups
soil, sand, water, various materials
    you choose

WHAT TO DO Think of all the things that a potato eye would need to grow properly. These might include food (part of the potato to which the eye is attached), water, soil, light, the right temperature, and others. You have as many choices for testing as there are eyes in the potato.

Cut the potato so each eye has about the same amount of potato attached to it.

Set up several containers filled with the same mixture of soil, sand, and water.

Plant the potato pieces so the eyes are just above the surface of the planting material. Provide one planted eye with *all* of the things you believe it needs to grow well. Deny *each* of the other planted potato eyes a different necessity. Keep all other conditions the same. Label each container so you know which necessity is being denied. Observe the eyes for several weeks to see if there are any differences in the way they grow. Keep a record in words and drawing of the plants' development.

Which *necessities* seem to be most important to the normal growth of the plant? What happens to the plants that are denied them?

birds for food. Against the light background, dark moths were easy targets for the birds. Light moths had a selective advantage in this environment. They were more likely to survive their hunters and more likely to produce offspring.

Industrialization produced a lot of soot and changed the background from light to dark. Against this new background, light-colored moths became the easy targets. Now the hunting birds were more likely to miss the darker moths, giving them a better chance of producing offspring. Within fifty years, this population of moths had changed from light to dark. The selectors were the hunting birds, with the pressure coming from the changed environment. This is just a small example of the kinds of forces that have shaped the living creatures sharing this planet.

The understanding of evolution is itself evolving. It brings together information from a wide number of scientific disciplines. Each of these fields is constantly making new discoveries. Each piece of information adds to the picture we have of the history of life on earth. The details will continue to change. Some things will not change. We know life has existed on this planet for a long time. We know that life forms have changed over the long history of the earth. Those that could not successfully adapt have been lost. We are beginning to appreciate that all organisms are related and depend on one another and their environment to survive. We can no longer afford to think of people as isolated from the environment, separated from the natural world. We are part of a community of living things.

BIBLIOGRAPHY

Asimov, Isaac. *How Did We Find Out about Our Genes?* New York: Walker, 1983.

Back, Christine, and Olesen Jens. *Chicken and Egg.* Morristown, N.J.: Silver Burdett, 1986.

Back, Christine, and Barrie Watts. *Spider's Web.* Morristown, N.J.: Silver Burdett, 1986.

Back, Christine, and Barrie Watts. *Tadpole and Frog.* Morristown, N.J.: Silver Burdett, 1986.

Bornstein, Jerry and Sandy. *What Is Genetics?* New York: Julian Messner, 1979.

Bornstein, Sandy and Jerry. *New Frontiers In Genetics.* New York: Julian Messner, 1984.

Gutnick, Martin. *Genetics: Projects for Young Scientists.* New York: Franklin Watts, 1985.

Higgins, Jane H. *Discovering Genetics.* East Aurora, N.Y.: DOK Publications, 1983.

Huxley, Julian. *The Wonderful World of Life.* Garden City, N.Y.: Doubleday, 1969.

Lerner, Marjorie Rush. *Who Do You Think You Are? The Story Of Heredity*. Englewood Cliffs, N.J.: Prentice-Hall, 1963.

Newell, Norman D. "Why Scientists Believe in Evolution," American Geological Institute pamphlet, 1984.

Nouvelle, Catherine, and Henri de Saint-Blanquat. *The Human Story: The First People*. Morristown, N.J.: Silver Burdett, 1986.

Showers, Paul. *Me and My Family Tree*. New York: Crowell, 1978.

Silverstein, Alvin and Virginia. *The Genetics Explosion*. New York: Winds Press, 1980.

Stwertka, Eve and Albert. *Genetic Engineering*. New York: Franklin Watts, 1982.

Taylor, Ron. *The Story of Evolution*. New York: Warwick Press, 1980.

Tiley, N. A. *Discovering DNA*. East Aurora, N.Y.: DOK Publications, 1983.

Wade, Nicholas. *The Ultimate Experiment: Man-Made Evolution*. New York: Walker, 1977.

Watts, Barrie. *Butterfly and Caterpillar*. Morristown, N.J.: Silver Burdett, 1985.

INDEX

A

A and B genes, 62–66
ACGT bases, 75–79, 85, 86
Adenine, 75
Algae, 98
Amino acids, 84–88
Amoebas, 26–28, 34, 35
Animal breeding, 7–10, 11
Animal cell, 22
Artificial selection, 12, 100

B

Babies, 23–24, 25, 26
Blood groups, 58–64
Blueprint, genetic. *See* DNA
Breeding, 7–12

C

Carbon, 84
Cell theory, 18
Cells
    as building blocks, 14–15
    daughter, 34, 71, 73
    discovery of, 17
    division, 26–34, 36, 50, 71,
      79, 80–81
    egg, 24–25, 26
    germ, 47, 48, 50
    invisible parts of, 32, 34
    membrane, 18, 19, 28, 30,
      31, 32, 79, 80
    nucleus, 19–21, 30–32, 34,
      48, 79, 80, 81, 82, 85
    parts of, 13–22
    plant, 21, 22, 30, 32

Cells *(cont.)*
    proteins in, 21, 74, 82, 84–89
    red, 26
    reduction division, 47–50
    resting stage, 80
    seen through microscope, 15–17, 94
    stored nutrients, 18, 19, 21
    strands, 30, 32, 50
    structure, 22
    walls, 21, 30
    See also Chromosomes; Cytoplasm
Characteristics. *See* Traits
Chicken egg, 15, 16
Chloroplasts, 21
Chromatin, 20
Chromosomes, 20, 30–32, 79, 80
    as genetic blueprint carriers, 73–74, 90
    reduction division, 47–50
Classification, organisms, 93
Cleft chin, 2, 60–61
Code, 85–86
Codominant genes, 63
Consumer organisms, 95
Cytoplasm, 18, 20, 21, 28, 34, 79, 80, 81, 82, 83, 84, 85
Cytosine, 75

D

Daughter cell, 34, 71, 73
Decomposer organisms, 95

Diseases, genetic, 58, 91
DNA (deoxyribonucleic acid)
    amino acid order and, 84–87
    bases, 75, 77–79, 81, 85, 86
    as blueprint, 70–75
    building blocks, 78
    handrail, 75, 76, 79
    mistakes, 88–91
    model, 75
    mRNA and, 82, 83, 85
    spiral staircase, 75, 76, 77, 79
Dogs, 8–10
Dominant trait, 40, 43, 45

E

Ear shape, 2
Earth, beginnings of, 95–96, 98–99
Egg, chicken, 15, 16
Egg, fertilized, 25–26, 50
Egg, frog, 25, 26, 36
Egg cell, 15, 16, 24–25, 26
Electron microscopes, 21
Energy, 13–14, 70, 95
Environment
    artificial selection, 100–101, 104
    changes affecting life forms, 91, 98
    effect on traits, 45, 47, 54
Enzyme, 86
Evolution, 95–104

Eyes
    color, 2, 10, 51, 58, 62, 64, 66
    shape, 23

F

Facial expression, 6, 23
Factors. *See* Genes
Family resemblances, 7, 9
Family traits, 52, 54
Family tree. *See* Pedigree
Fertilized egg, 25–26, 50
Fibrin, 84
Fixed materials, 29
Fossils, 95–96, 97
Frog eggs, 25, 26, 36

G

Gametes, 47, 48, 50
Generations, 55
Genes
    A and B, 62–66
    in DNA, 82, 83, 90
    dominant, 40, 43, 45
    and environment, 45, 47, 54
    genotype, 52–53, 61
    identical, 102–103
    O, 62, 63
    pairs, 47, 62, 64
    and phenotype, 51–52
    plant, 43, 46, 47, 48, 50
    recessive, 41, 43
    and traits, 10–11, 43, 45–47
    See also Pedigree

Genetic blueprint. *See* DNA
Genetic diseases, 58, 91
Genetics, 4, 47, 50, 51, 58
Genotype, 52–53, 54, 61
Geologists, 95
Germ cell, 47, 48, 50
Greeks, ancient, 10
Green plants. *See* Plants
Growth hormone, 84
Guanine, 75

H

Hair, 2, 6–7, 54, 66
Hand and foot size, 2
Height, 2, 10, 45, 51, 54
Hooke, Robert, 17
Hormones, 84
Human traits, 5–7, 60
Hybrids, 40, 42, 43–45, 52, 53
Hydra, 36–37
Hydrogen, 84

I

Identical genes, 102–103
Identical twins, 6
Industrialization, 104
Inheritance, 4
    of disease, 58, 91
    human, 52, 54, 60
    Mendel's theory, 45
    paired chromosomes, 47–48
Interviewing. *See* Pedigree

L

Lenses, 15
Livestock breeding, 11
Lizard reproduction, 36

M

Mannerisms, 2
Meiosis. *See* Reduction division
Membrane, cell, 18, 19, 28, 30, 31, 32, 79, 80
Mendel, Gregor, 38–48, 52, 73
Microscopes, 15, 17, 21, 28, 29–32, 74, 94
Mitosis, 26–34, 36, 50, 71, 79, 80–81
Model, DNA, 75
Molecule, DNA. *See* DNA
Molecule, RNA. *See* RNA
Molecules, protein. *See* Proteins
Morning glory, hybrid color, 45
Moths, artificial selection, 101, 104
mRNA chain, 82–83, 85–88, 94

N

Nitrogen, 84
Nose shape, 10
Nucleus
    in cell division, 30–32
    as control center, 19–21, 34
    DNA in, 79, 81, 82, 85
    and reduction division, 48
    in resting cell, 80

Nutrients
    cell, 18, 19, 21
    energy and, 95
    plant, 30

O

O gene, 62, 63
Offspring, 8, 10, 35, 38, 40, 41, 42–48, 52–53
Organisms
    amoeba, 26–28
    classification, 93
    connections between, 92–104
    definition, 14
    fixed materials, 29
    identical, 102–103
    similarities and differences, 95
    survival tasks, 72
Oxygen, 84

P

Paleontologists, 95–96
Parents, 36–38, 40, 47–48, 50, 90. *See also* Offspring; Reproduction
Pea plants, 38–50, 52, 59
Pedigree
    blood types, 62–63
    building your own, 67–69
    drawing, 54–57
    eye color, 64, 66

relatives included, 57–58
trait record, 58–62, 69
Phenotype, 51, 58
Pistils, 40
Plants
    breeding programs, 11–12
    cell wall, 21, 30
    cells, 21, 22, 30, 32
    chloroplasts, 21
    genotype, 52–53
    hybrids, 40, 42, 43–45, 52,
      53
    nutrients, 30
    offspring, 40–43, 46
    roots, 30, 32
    seeds, 40, 50, 52
Pollen, 40
Ponds, 27, 92–93
Potatoes, 102–103
Producer organisms, 95
Proteins, 21, 74, 82, 84–86, 88–
  89
Pure traits, 40, 41

R

Recessive traits, 40, 42, 43
Reduction division, 47–50
Relatives. *See* Family; Parents;
  Pedigree; Siblings
Reproduction, 27, 35–38, 50. *See
  also* Offspring
RNA (ribonucleic acid), 82–83,
  85–88, 94
Rocks, studies of, 95, 96, 98

Roots, plant, 30, 32

S

Seeds, 40, 50, 52
Sequoia trees, 35
Sexual reproduction, 50
Siblings, 55
Skin, 2, 64
Still photos, 29–32
Sun, 95, 96
Survival tasks, 72

T

Talents, 2
Thymine, 75
Traits
    animals and plants, 8, 10,
      11–12
    definition, 2
    dominant, 40, 43, 45
    family, 2–4, 50–54, 58–62,
      69
    genes influencing, 10–11,
      43, 45–47, 54, 59, 61
    human, 5–7, 60
    identical twins', 6
    parental, 40, 47–48
    pure, 40, 41
    recessive, 40, 42, 43
    results of environment, 54,
      100, 101, 104
    See also Pea plants
Twins, identical, 6

U

U base, 82, 86

V

Voices, 6
Volcanic activity, 96

W

Weight, 2, 54

ABOUT THE AUTHOR

Sandy Bornstein graduated from Barnard College and holds a master's degree in human genetics from McGill University. She has worked in the medical genetics program at Brookdale Medical Center in New York City and at the New York University-Bellevue Medical Center Cytogenetics Laboratory. She left the genetics laboratory to join the laboratory of the elementary school classroom. She now teaches science in a small independent school. Ms. Bornstein, who is married and the mother of two daughters, currently lives in Brooklyn, New York. Her two previous books are *What Is Genetics?* and *New Frontiers in Genetics*, both written with her husband Jerry.